Hydrogen

Facing the energy challenges of the 21st century

Canadian Hydrogen Association

Association Française de l'Hydrogène

Tapan Bose and Pierre Malbrunot

with the collaboration of Pierre Bénard and John Viola

ISBN : 978-2-74-200639-7

We would like to thank our sponsors, namely: Natural Resources of Canada, the Canadian Hydrogen Association, the University of Toronto and the Université du Québec à Trois-Rivières for their help in the realization of this book. We would also like to thank the French Minister of the National Education for Graduate Studies and Research for his interest in the book.

We are grateful to Thierry Alleau, Stephen Boucher, Richard Chahine, Raymond Courteau, Claude Derive, Kamiel Gabriel, Ted Hollinger, Ron Venter, and Mrs. Lucie Bellemare for their assistance and encouragement.

Tapan Bose
Pierre Malbrunot
Pierre Bénard
John Viola

A French version of this book written by the same authors was published in Dec. 2006 by John Libbey.

Editions John Libbey Eurotext
127, avenue de la République, 92120 Montrouge, France
Tél : 33 (0) 1 46 73 06 60 ; Fax : 33 (0) 1 01 40 84 09 99
Email : contact@jle.com
Site internet : http://www.jle.com
Éditrice : Raphaèle Dorniol

John Libbey Eurotext Limited
42 - 46 High Street
Esher
KT109QY
United Kingdom

Preface

It gives us great pleasure to sponsor this book on hydrogen energy technologies by the Canadian Hydrogen Association and the Association Française de l'Hydrogène.

Canada is committed to a clean environment and to reducing greenhouse gas emissions. In this context of increasing concern about these emissions, this book is certainly very timely. Amply illustrated, it addresses the issues and details of all aspects of the use of hydrogen as an energy vector, a supplement and possible replacement of fossil fuels. The book is available in both the official languages of Canada: English and French.

Natural Resources Canada has supported hydrogen energy technologies in Canada dating back to the 1980s. Its support has created the leadership that Canada now enjoys in the world.

The Canadian Hydrogen Association is the voice of the hydrogen community in Canada and is a trusted advisor to both government and industry.

The Centre for Hydrogen & Electrochemical Studies (CHES) of the University of Toronto and the Hydrogen Research Institute (HRI) of the Université du Québec à Trois-Rivières have both distinguished themselves through long-standing research program in hydrogen energy technologies.

We hope that you enjoy this book and learn more about hydrogen energy technologies.

Nick Beck
Natural Resources Canada

Alexander Stuart
Canadian Hydrogen Association

Ronald Venter
Centre for Hydrogen
& Electrochemical Studies
University of Toronto

René-Paul Fournier
Université du Québec à Trois-Rivières

Preface

(translated from the french preface)

We have entered an era of dwindling and expensive oil, it is an observation. The increased use of fossil fuel is changing our climate. That is a certainty.

The time has come to develop alternatives to fossil energies. It is now a necessity.

This is why France spends each year more than one billion euros for research to find an energy solution after oil in technologies such as:

Bioenergies, biofuels bioreactors for fuel synthesis;
Nuclear energy, including technologies for the production of hydrogen;
Thermonuclear fusion, the *energy of stars*, with the international project ITER that France has impamented in Cadarache, France;
And finally, hydrogen and fuel cells.

This last technology already shows astonishing and promising prospects. The young people of tomorrow will charge their cellular telephones with a hydrogen cartridge. Everyone will have at home what it takes to produce hydrogen from solar energy to supply their car with water as the only emission. It is not science fiction anymore; it is already a reality in laboratories today!

These perspectives may cause within the span of one generation profound changes in our society, our habits, our work and consequently our behavior. We should be prepared.

I am particularly happy to write the preface to this collective work of the AFH2 and its Canadian counterpart. This important book presents in a very clear way science and technologies of the hydrogen economy. It is for everyone, young and old who want to learn, understand and anticipate the advances to come.

Gilles de Robien
Minister of French National Education for
Graduate Studies and Research

Table of content

The hare, the tortoise and hydrogen

A hare and a tortoise[1] were discussing Nature, whose future looked bleak because of the activities of Man.
To save the Earth, they were told, one could use Hydrogen to fuel a world starving for energy.
In order to learn more, the two friends began to look for information. They eventually stumbled on this book, which answered their questions while respecting their pace.
The pages on the right provide a brief but richly illustrated topical discussion, adapted to the needs of an agile runner, who, less interested in details, would rather learn from the illustrations.
The pages on the left delve into more detail, addressing the needs of a patient and systematic reader who wants to learn more.
The hare runs forwards and backwards, jumping merrily from page to page, while the tortoise patiently and methodically goes through the texts, one chapter after the other.
How does this fable end?
They both learn in their own way what they wanted to know.

Basic information on energy can be found at the end of this book.

The main scientific and technical terms, as well as the units used throughout this book are explained as they are introduced in the main text. Their definition is also provided at the end in a glossary, which, in addition, contains a list of abbreviations, acronyms and symbols. The numerical values of various constants and properties given throughout the text are not intended to be used as reference. They rather represent typical, "order of magnitude" estimates whose sole purpose is to illustrate the points discussed in the book. This book should in no way be considered as a reference for the values of physical constants and other properties.

[1] Maybe the same pair as in La Fontaine's and Aesop's Fables?

« Ενεργια », force in action...

In our daily lives, everything we do involves:

Movement, bringing motion to objects at rest

From cars to planes to ships to missiles – overcoming friction from ground and air;

Overcoming heat and cold

By warming our homes, schools and offices; by feeding ourselves, because our body heat comes from the food we eat.

Lighting

Provided by a candle, an electric bulb or the sun.

Production of food

Sowing seeds, growing crops, and harvesting crops.

Manufacturing and transforming

Melting and manufacturing metals, producing and molding plastics, transforming and finishing wood, grinding cereals and baking bread.

Construction

Excavating the ground, preparing and pouring concrete, hoisting and assembling various materials for structures.

Communication

By telephone, radio, and television, Internet or simply through speech, using our voices.

And so much more!

All of these actions would not be possible without energy*[1], an abstract notion difficult to define; usually associated with the ability to perform work. The word itself comes from the Greek word (energia) which means "force in action".

How many of the following does a typical family use?

- ● ● ● Combustion engine, for instance in cars, lawnmowers, two (or more) -wheelers
- ● ● ● Electrical engines, for instance in cars, appliances, toys, multimedia equipment, gardening, cooking
- ● ● ● Batteries and cells, for instance in toys, cell phones, digital cameras, flashlights, remote controls, watches, computers, etc

Answers:

(a) from 1 to 6. (b) from 25 to more than 70.
(c) from 15 to more than 50!

[1]The asterisks refer to the glossary at the end of this book.

...Energy is everywhere!

From well to wheel...

Primary Sources of energy*

Energy is lost each time it changes form, until it is eventually converted into low temperature heat* of little practical value, that eventually dissipates into the environment. In order to perform work, energy must be continuously generated from a primary source.
From the dawn of history we have performed work using our own muscles and through domestic animals – all requiring a primary source of energy, supplied by food. Flowing water and wind were eventually harnessed, increasing our capacity to perform work and providing some relief from the chores needed for survival. The combustion of wood was also used as a primary source of energy, for heating and cooking. With the dawning of the industrial age, coal, in ever greater amounts, was used to quench our thirst for energy. This was followed by increasingly efficient fuels like oil, and other primary sources such as hydraulic dams.
Four types of primary energy sources are in use today:
Fossil fuels* – coal, oil and natural gas – represent about 90% of world energy consumption. By definition, their reserve is not inexhaustible, and the available stock ranges from a few centuries for coal to a matter of decades for oil and natural gas. These fuels need to be refined before they can be used in their final applications.
Renewable energy* originates more or less directly from the sun. Hydraulic dams, wind turbines and photoelectric solar panels supply us with electricity. Crops can be turned into fuel and solar radiation can be used for heating water at low temperatures. By their very nature, the sources of renewable energy are inexhaustible on timescales relevant to human beings, yet they face two problems: they are intermittent and they need to be stored to manage supply and demand; in addition, they are rather diffuse and as such, they require large investments.
Geothermal energy*, similar to renewable energy, it is obtained by capturing some of the heat released by the core of the earth in locations where magma is close to the surface.
Nuclear energy* is currently obtained through the fission of uranium atoms in reactors from which large quantities of heat are produced. This is used to generate steam which drives turbines, as in classical thermal plants.

Storage, transportation and distribution

Energy must be stored in order to meet the requirements of supply and demand. Fossil fuels such as coal, oil and natural gas can store a large amount of energy in very compact form.
Primary energy sources are seldom used at their production site. They must be transported on a large scale, often over long distances by sea, by land or by train and, in the case of liquids and gases, by pipeline.
A significant fraction of a primary source of energy is required for its transportation and distribution.
In general, the energy required to transport and distribute primary source of energy makes up a significant fraction of the energy delivered and hence its cost.

Electricity as an energy vector

Although electricity occurs in nature as lightning, it cannot be used in this form to perform any practical work. To be of any use to us it must be produced at a more predictable rate through the transformation of mechanical energy in power plants* from a primary source. Thus, electricity constitutes a secondary source of energy –or an energy vector- although its presence everywhere in our daily lives may give us the impression that it is a primary source of energy. Electricity is most often produced by large scale thermal plants* using coal or nuclear fissile matter as fuel, or by hydroelectric power stations, and less frequently by power plants using wind, solar (photovoltaic*), geothermal or tidal energy. Once generated, electricity must be transported and distributed because, although it can be stored in accumulators in small quantities for specific applications, it cannot be stored efficiently in large quantities, and must be used as soon as it is produced. Electricity is **thus an energy vector, and not a source of energy.**

...The flow of energy

Fuels such as oil, gas and coal are primary sources of energy that can be obtained directly from oilfields or coal mines. Nuclear fuel is a non-fossil primary source of energy obtained from uranium deposits.
Rivers, tides, and wind generate mechanical energy that can be harnessed and converted, as can solar radiation, into heat or electricity. These energy sources are renewable and will not vanish on timescales relevant to human experience.
Electricity must, however, be produced from a primary source. Although it can be easily transported, it cannot be stored in large quantities, and must be regarded as an energy vector, instead of an energy source.

Energy...

Who would accept today living without electricity or without their car? Could we agree to go back to a time when homes were poorly heated and the streets were not lighted at night?

Civilization requires energy. Progress has required access to increasing levels of energy. It began with the discovery and the mastery of fire by prehistoric men, initially for protection, heating and lighting; and later to produce pottery and metallic objects. Mankind learned to rely on the work provided by animals and slaves, and eventually to produce energy from wind and water. With the advent of the industrial revolution, our level of energy consumption rose dramatically. After relying for thousands of years on wood, we began to use coal, then oil and eventually electricity. The consequence was an unprecedented rate of economic development and substantial improvement in our standard of living, an increased level of exchanges, a more efficient industrial infrastructure, freedom from the most arduous tasks, more comfort and a general improvement of health.

Energy as an engine of prosperity

There is a close correlation between standard of living and level of energy use. In North America, every person uses 6.6 ton equivalent petroleum per year (tep), compared to 3.4 on average in Europe and 0.6 for a person living in Africa.
Differences in energy consumption are possible among countries with similar standards of living (such as Canada and the United States for climatic reasons). However, it is estimated that to attain an economic development comparable to the Western World requires an energy consumption of over 2 tep per year per person.

Economic zone	Energy use 1999 (ton equivalent petroleum (tep) per year per person)
World average	**1.7**
North America	**6.5**
USA	8.1
Latin America	**1.1**
Western Europe	**3.4**
Germany	4.2
France	4.2
Italy	2.9
Africa	**0.6**
Middle East	**2.3**
Far East	**0.9**
China	0.9
Japan	4.1

Ref.: Minister of Finance and Industry of France, Paris 2004

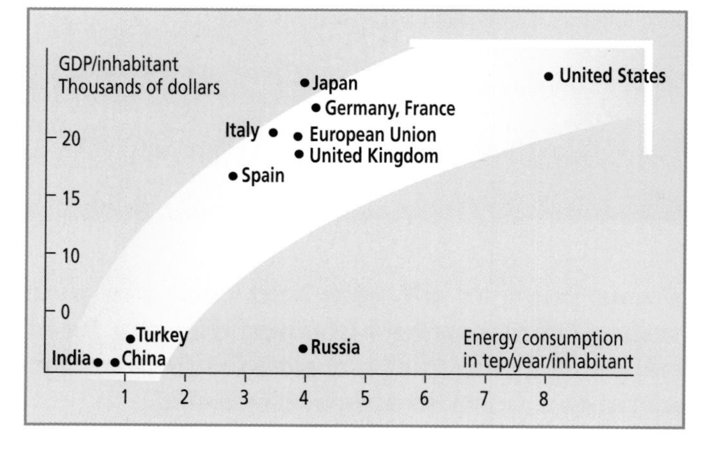

Source : ibid.

...as an agent of progress

During the 19th and 20th centuries, man has progressively freed himself from physical labor and reduced his use of animals by inventing the steam engine and the internal combustion engine. This was followed by the discovery of electricity: electronics and computers. These technologies caused deep economic and societal changes. Mechanized agriculture, automation, fast, safe and comfortable transportation systems, quasi-instantaneous communication, new scientific and medical breakthroughs, space as the new frontier are all examples of human progress and the consequence of increased use of energy in all of its forms.

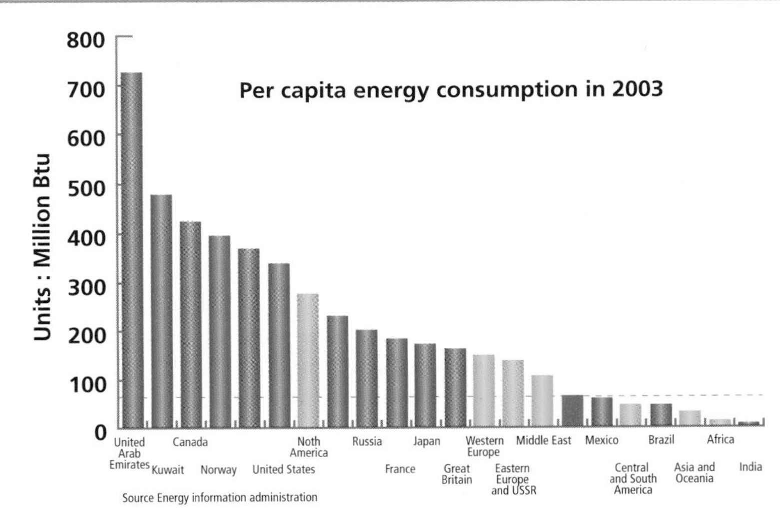

To irrigate a plantation, this pouch of 50 liters was raised by a camel from a depth of 20 meters in one minute. The same work could have been done by a pump using 200 watts of electricity supplied by 2 sq. meters of photovoltaic cells.

Using too much energy...

The world's energy consumption takes off... and its distribution changes

The benefits of energy use, together with an increasing population worldwide, have lead to a rapid rise in the demand for energy. During the second half of the 20th century, the number of people on the planet has increased from 2.5 to 6 billion, and energy consumption has more than quadrupled. According to the United Nations, the total population should stabilize to a level of about 9 billion people in 2050. By then, energy consumption will have doubled. The emerging economies would then represent about 60 to 70% of total energy consumption per year, compared to 10 to 15% today.

Excessive exploitation of fossil fuels leads to serious problems

Currently, fossil fuels such as coal, oil and gas satisfy about 80% of the world's energy requirements. Renewable energies (mainly through biomass and hydroelectricity) represent less than 14% of the total, and nuclear energy accounts for the rest. At our current rate of consumption, humanity will have exhausted most fossil fuels in about two centuries: fuels that took nature hundreds of millions of years to produce. Serious undesirable side-effects of our use of fossil fuels are beginning to appear, and they are expected to intensify if we continue on our present course.

A first negative consequence of the excessive use of fossil fuels is climate change. On a global scale, the combustion of carbon-rich fossil fuels releases carbon dioxide (CO_2) in the air, increasing its atmospheric concentration. This results in an intensified greenhouse effect and in climate change. Pressure from the public has resulted in an international agreement, the Kyoto protocol, which seeks to lower the greenhouse gases emitted by energy production, heating, transportation, industrial processes and agriculture by the year 2012.

The greenhouse effect is a process that causes warming of the earth through the action of certain atmospheric gases on the incident solar radiation – in particular water vapor, carbon dioxide, methane, nitrous oxide, ozone and chlorofluorocarbons. These gases, the so-called "greenhouse gases", act in the same way as the glass panes in a greenhouse. They let through short wavelength solar radiation, which can then be absorbed by the earth and re-emitted as long wavelength infrared radiation. The greenhouse gases of the atmosphere then absorb this infrared radiation, resulting in a net increase of its temperature. This natural phenomenon is the reason for our temperate climate. It is essential to life on earth as we know it because it raises the temperature of the atmosphere by about 30 °C above what it would be in its absence. However, since the beginning of the Industrial Revolution, the atmospheric concentration of certain greenhouse gases, mainly carbon dioxide and methane, has increased dramatically due to human activities such as the combustion of fossil fuels, the intensification of agricultural activities and deforestation. This has resulted in an additional artificial greenhouse effect – above and beyond the natural greenhouse effect- which is likely the cause of the global warming observed since the beginning of the 20th century (especially since 1960).

Another consequence of excessive use of fossil fuels is depletion. Fossil fuels have been produced as a result of complex natural processes during millions of years. At the current rate of consumption, the known reserves of fossil fuels such as oil and natural gas will only last for a few decades. The known reserves of coal are expected to last longer (a few centuries). The price of fossil fuel will naturally rise as they become less available. The uneven distribution of fossil hydrocarbons worldwide has had significant political and economic consequences. Geopolitics has encouraged several countries to try to lower their dependence on foreign sources of fossil fuels. France has lowered its dependence from 80% to 50% between 1973 to 2000 by using nuclear energy. Canada, as an exporter of energy, produces 150% of its energy needs.

In addition, burning fossil fuels causes pollution at a local level. The use of fossil fuels results in the local emission of pollutants which are chemically very active. Ozone (O_3) is caused by the interaction of solar radiation with nitrogen oxides (NOx) and volatile organic compounds (VOC). In the lower atmosphere, ozone affects the lungs and the eyes. It can also affect agricul-

Fuel	Proven reserves Gigaton equivalent petroleum (Gtep)	World consumption in 2000 (Gtep/year)	Proven reserves (years) (at current rate of use)
Oil	≈ 140	3.6	40
Natural Gas	≈ 130	2.2	60
Coal	≈ 500	2.2	> 200

tural production and constitutes a contributing factor in the formation of acid rain. The combination of nitrogen and oxygen during high temperature combustion processes results in the production of NO_x. Volatile organic compounds occur as a result of incomplete combustion of fossil fuels. They are also emitted during refueling.

Other harmful emissions from fossil fuels include particulate matter, toxic metals and polycyclic hydrocarbons, all of which can cause cancer. Particulate matter is produced by diesel engines. Carbon monoxide (CO) is also a by-product of the combustion of fossil fuels. Carbon monoxide is deadly in conditions of prolonged exposure and contributes to ozone formation. These emissions represent a serious public health issue, particularly in urban environments. In addition, the noise level produced by car engines has an impact on the quality of life.

...puts our environment at risk!

Our massive use of energy has only been possible by extensively tapping into the natural reserves of fossil fuels (coal, oil and natural gas) and of uranium. With limited quantities available, we are at risk of exhausting these resources within a time frame ranging from a few decades to centuries.

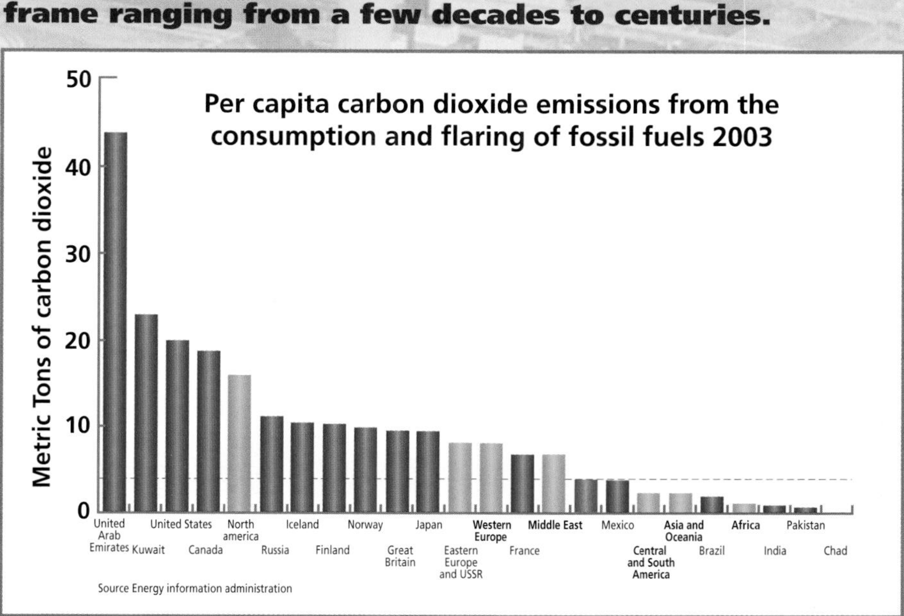

Worse, this consumption has serious ramifications on the environment and on public health:

- It increases dangerously the greenhouse effect by emitting carbon dioxide[1], methane and nitrous oxide, thereby causing global warming and other climatic perturbations.
- It depletes non renewable primary energy sources.
- It pollutes the environment by the release of toxic gases (nitrogen dioxide, ozone, aromatic compounds) that adversely affect public health and the environment.
- It affects the urban environment by contributing to smog through the emission of ozone and particulate matter, which are also harmful to our health
- It produces long-lasting nuclear waste which must be managed carefully.

[1]Since the beginning of the industrial era, carbon dioxide concentration has increased by 50% and it could double by the end of the century.

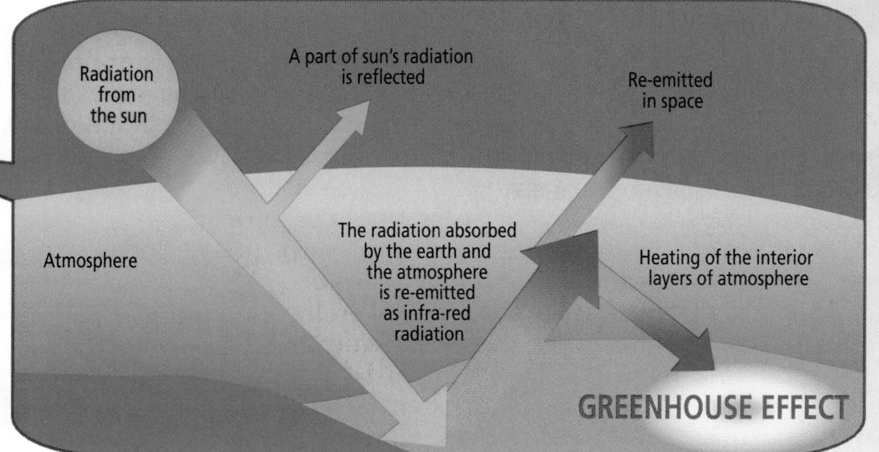

Possible solutions...

Could we consider for a moment giving up the benefits that energy provides? Can we limit the access of the greater part of the world's population to energy, considering that most people on Earth have yet to benefit from them?

Consume less and more efficiently

About half the energy consumed by countries such as Canada and France is used to satisfy the energy demands of residential and tertiary buildings. Of this amount, two thirds are devoted to warming water for domestic use and for heat production. Since the first oil crisis, regulations and improved thermal insulation have resulted in a decrease by a factor of three of this energy expenditure for new constructions. Cogeneration, which consists in recovering unused low temperature heat emitted by thermal power plants, is a useful way of increasing energy efficiency. This process, even though it results in lowered efficiency from the point of view of strict production of electricity, leads to a substantial improvement of the overall efficiency of power plants. It could be used in densely populated urban areas where a heating infrastructure could be implemented. In addition, constant progress is being made toward increasing the efficiency and the level of temperature control of domestic heating systems. Better yet, heat pumps, which can transfer external heat to a building, also represent an interesting possibility toward lowering energy consumption.
The transportation sector is responsible for 30% of anthropogenic emissions of CO_2 in developed countries, and about 12 to 15% of emissions worldwide. Automobile manufacturers seek to improve the energy efficiency of vehicles through new engine designs, improved aerodynamics, better tires and the use of lighter materials. They are, however, constrained by consumer demands on things such as vehicle size and noise reduction. In Europe, the car industry has promised to lower CO_2 emissions to 140 grams of CO_2 per km in 2008 and 120 in 2012, compared to 190 in 1999.

Use fossil fuels more efficiently

Various scenarios predict that the contribution of fossil fuels could be reduced to about 60% of the total energy consumption by 2050. Considering the increased energy demand worldwide, in absolute numbers this would still represent an increase of 50 to 100% compared to the current level. New sources of oil, however small, will be sought, and their extraction rate, which has increased from 35% to 45% over the last 30 years, will be further increased.
Coal, which accounts for 23.5% of world energy consumption, will continue to play a significant role because it is relatively abundant and well distributed worldwide. As oil supplies dwindle, coal could be considered a viable alternative to oil in the United States, India and China, who currently use it in large thermal power plants to produce electricity, provided toxic emissions are reduced. However, coal is responsible for 30% of the global emission of CO_2.
A significant shift towards using natural gas is currently underway. The reserves of natural gas are considered to be larger than oil. In addition, natural gas emits 40% less CO_2 than oil, because methane, which constitutes 85% of natural gas, contains twice as many hydrogen atoms per carbon atom. Leaks of natural gas, however, must be prevented because methane causes a greenhouse effect 23 times greater than CO_2.
Even if the risks of depletion could be alleviated, CO_2 emissions would still remain problematic. In addition to natural CO_2 sinks (such as vegetation), whose capacity could be increased by extending the forest cover, artificial means to sequester CO_2 at its point of origin have been the subject of intense interest recently. Carbon sequestration would require collecting it at its point of emission, and sequestering it in natural cavities in the earth's crust where it would remain confined. This would only be practical for large scale installations such as power plants or synthetic fuel production facilities.

Use Biomass*

Biomass already represents about 10% of world energy consumption, essentially because of the conventional use of wood as a fuel. Biomass is termed CO_2 neutral, meaning that it does not contribute to the increase of carbon dioxide concentration in the atmosphere, because the CO_2 emitted when energy is extracted from biomass is recaptured during its subsequent production. This neutral effect of biomass on the atmospheric concentration of carbon dioxide requires that biomass be regrown. Biomass offers a steady supply of energy. The large-scale production of biomass would require addressing several issues, such as the availability of land for biomass production, overall efficiencies, the consequences of intensive agriculture, pollution due to extensive use of fertilizers, etc.

...A wiser use of our current energy sources

In order to avoid the depletion of fossil energy sources, to reduce greenhouse gas emissions and pollution in general, we should at the same time:

- **Reduce energy consumption by rationalizing its use.**

- **Use environmentally friendlier fuels with a lower CO_2 footprint, especially in the context of the Kyoto agreement, which has been adopted by most countries around the world, and by which they agree to lower their CO_2 emissions. For example, the use of biomass as an energy source (to produce green fuels, for instance) respects the spirit of the Kyoto agreement.**

Another solution…

Renewable energy sources

These energies ultimately all come more or less directly, from the sun. They do not produce greenhouse gases[1]. They will not affect the overall equilibrium of our planet. They are abundant, and, on a human scale at least, they can be considered inexhaustible (the sun will last for 4 to 5 billion years more). However, renewable energies are inherently transient in nature. Their production rate varies on a daily and seasonal basis. They are subject to weather patterns. Therefore, these energy sources usually require some form of storage to balance supply and demand. In addition, their energy density is generally low, and they require large scale facilities to harvest them.

Hydraulic power is the most commonly used renewable energy source to produce electricity. It can store energy through the water reservoirs created by dams, and can produce electricity on a continuous basis. Hydraulic power requires an important initial capital investment, which can be recovered over a long time, so that the production cost per kWh of energy can actually be quite low. Developed countries have harnessed almost all the available sites for dams. However, a significant number of sites remain open for development in emerging economies. The drawback of hydraulic power is the large footprint of the infrastructure needed, which can require population displacement and may provoke dramatic environmental changes at a local level. It can also adversely affect the habitat of certain species.

Wind turbines, the modern version of windmills, can now competitively produce electricity. One wind turbine can produce up to 3 to 5 Megawatts of electricity. Their drawback is the intermittent nature of wind power, because of which they generate electricity only 30% of the time on average. If the wind turbine is isolated and not connected to a grid, some form of energy storage or a second primary energy source is required to regulate supply and demand.

Photovoltaic solar cells can directly convert solar radiation to electricity with an efficiency of about 10%. Their cost has come down over the last 30 years, but they still remain expensive. They have a negligible impact on current energy production but they have a strong potential for growth, particularly in isolated areas.

Thermal solar panels can be used to produce clean warm water with an efficiency of 40 to 50%. They can be used for heat generation in cold and sunny locations. They can also be used to concentrate solar radiation to generate temperatures of about a few hundred degrees for certain industrial processes.

Geothermal energy is based on the concept of using the heat generated from the earth's core. It is inherently local, in the sense that it can only be harnessed where the earth's crust is relatively thin, such as in Iceland and Italy, and cannot be considered for energy production on a large scale. As we define it, geothermal energy does not include the heat from a few meters below ground level used by small thermal pumps for domestic applications.

Why not nuclear energy?

Nuclear energy generally provokes fear and distrust. Based on splitting unstable atoms, it is a mysterious and complex field, difficult to grasp without the appropriate scientific background[2]. When thinking about nuclear energy, people think of Chernobyl in the Ukraine and of Three Mile Island in the United States. Beyond the bad press, an objective look at Chernobyl reveals that it would never have happened if proper safety protocols had been in effect. In addition, the Three Mile Island accident was caused by human error and had essentially no impact on the public besides contributing to the cloud of distrust lingering over nuclear energy. People are weary of managing nuclear waste, some of which has a long lifetime. There are, however, answers to this issue, and considerable progress has been made toward a safe and satisfactory solution. Nuclear energy offers the remarkable advantage of large scale production of electricity without emitting CO_2. Today, 440 nuclear power plants produce about 7% of the world's energy requirements, and 17% of its electricity. If nuclear power was the sole source of electricity worldwide, CO_2 emissions would drop by 25% to 30%.

Research and development activities in this field have been directed toward increasing safety; reducing and eliminating long-lasting radioactive waste; lowering the likelihood of unintended military uses of nuclear waste; developing more efficient high temperature reactors and super generators, which would eliminate all chances of eventually running out of nuclear fuel.
Longer term research and development activities are geared towards achieving nuclear fusion, which, with a low yield of nuclear waste, could produce more energy than fission and whose fuels (deuterium and tritium) are available in large quantities.

[1]A small amount is produced if the equipment necessary to harness renewable energy sources is manufactured using fossil fuels.
[2]See, for instance, in the Annex "Other forms of energy".
[3]Ibid.

...Carbon-free energy

It would be possible to avoid carbon dioxide emissions by:

- **Using renewable energy sources, such as hydraulic energy (hydroelectric power plants), solar energy (photovoltaic solar cells and thermal solar panels), wind power (wind turbines) or geothermal energy (heat from the earth's core).**

- **Pursuing and intensifying our current use of nuclear power using fission while hoping that fusion[3] becomes viable one day.**

A promising solution...

The hydrogen atom and the hydrogen molecule

Hydrogen (represented by the symbol H) is the simplest and lightest of all chemical elements. Its atom is composed of a single positively charged proton – the nucleus - around which revolves a single negatively charged electron. Present at the beginning of the Universe, hydrogen is the ancestor of all other heavier and more complex elements, ultimately formed by fusion processes involving hydrogen nuclei. In the sun, hydrogen is the fuel from which solar radiation is emitted, through a fusion reaction that produces helium. Most of our energy sources ultimately come from this process. Hydrogen does not exist on earth in its atomic form, it is only found combined with other elements in molecules (most frequently as water). Hydrogen can be extracted from several compounds (such as water) to produce molecular hydrogen gas which is a very light (fourteen times lighter than air). Like all gases, hydrogen can be liquefied if cooled sufficiently, but only at the very low temperature of -253 °C. Hydrogen can become solid at - 259 °C.

High energy, but low density

Hydrogen is a fuel with high energy content per unit of mass. When it is combined with oxygen during combustion, producing only water as a by-product, it releases 120 Megajoules of energy per kilogram of hydrogen (33.3 kWh/kg), not including an amount of 20 MJ/kg contained in the residual water vapour. To obtain a similar amount of energy would require 2.5 kg of natural gas, 2.75 kg of oil or between 3.7 to 4.5 kg of coal. On the other hand, hydrogen is a very light gas, and even though it has the highest energy content of all fuels per unit of mass, it has one of the lowest when the energy content is measured per unit of volume. Its density at ambient pressure and tempe-rature is only 90 grams per cubic meter (0.09 kg/m^3), 8 times lower than natural gas! In order to produce the same amount of energy from a given volume of natural gas, a volume of hydrogen two and a half times greater is required under ambient conditions. The density of liquid hydrogen is also very low (70.8 kg/m^3): one cubic meter of water weighs 1 ton compared to 70.8 kg for one cubic meter of liquid hydrogen. Thus, the energy content of a litre of liquid hydrogen corresponds to the energy content of only 0.27 litre of gasoline (which is ten times denser than liquid hydrogen). As a consequence, hydrogen requires large volumes for storage, which can be problematic for certain applications.

Hydrogen, the fuel of the 21st century?

Whatever its source, primary energy must reach users through an efficient energy vector, which can minimize its adverse effects and can be used in a wide range of applications. Hydrogen could be this vector, because it can be produced from any primary or renewable energy source. In addition, it can be stored and transported to end-users and, in combination with oxygen, restores, when needed, the energy used to produce it without pollution.

Hydrogen properties	Values	
Atomic mass	1.0079	
Density at a pressure of one atmosphere and 273 K	0.09	kgm-3
Density of the vapor at a pressure of one atmosphere and 20.3 K	1.34	kgm-3
Density of the liquid at a pressure of one atmosphere and 20.3 K	70.79	kg.m-3
Lower heating value	119,900	kJ.kg-1
Higher heating value	141,800	kJ.kg-1
Heat of vaporization at 20.3 K	445.4	kJ.kg-1
Theoretical energy of condensation	14,112 (3.92 kWh.kg-1)	J.g-1
Temperature of fusion (at 1013 mbar abs.)	14.01	K
Temperature of vaporization (at 1013 mbar abs.)	20.268	K

...hydrogen!

...a carbon-free fuel[1]
...a clean and storable energy vector for better use of renewable resources

Comparative table of the properties of hydrogen relevant to its use as an energy vector with other fuels.

	Units	H_2 gas	Natural gas	Gasoline	Oil	Methanol
Density	kg/m³	0.090 (gas) *71 (liquid)*	0,721	738	840	787
Lower heating value (does not include the condensation energy of water)	Mjoule/kg	120	50	47	39	20
	MJoule/litre	0.01 (gas) *8,52 (liquid)*	0.036	34.6	32.7	15.4
Gasoline equivalence	litre	3 200 (gas) *4.06 (liquid)*	961	1	1.06	2.1
Natural gas equivalence	m³	3.3	1	#0.001	#0.001	#0.002

At room temperature, hydrogen is a gas made up of 75% of ortho-hydrogen and 25% of para-hydrogen. These two forms of hydrogen molecule differ from one another by the orientation of the spin of their proton. The spin of a proton is its kinetic moment associated with its rotation. The spins of the two protons of an ortho-hydrogen molecule are parallel. They point in opposite directions for para-hydrogen. The proportion of ortho and para in hydrogen gas is a function of temperature.
For instance, liquefied hydrogen is made of 99.8% para-hydrogen. Changing from one form of hydrogen to another is a reversible thermodynamic process. The process is exothermic when ortho-hydrogen is converted to para-hydrogen. This implies that liquefying hydrogen requires an additional amount of energy, above and beyond the amount required for cooling and absorbing the latent heat of liquefaction. The transformation is endothermic in the opposite direction (para to ortho).

Gaseous hydrogen is "molecular", formed by two hydrogen atoms. It is represented by the symbol H_2. The combustion of hydrogen, a process during which it combines with oxygen, produces a significant amount of energy per unit mass with only water as a by-product. Thus, hydrogen is a high quality fuel that can generate heat and mechanical energy without the emission of pollutants or greenhouse gases. Hydrogen has another important property: it can be combined with oxygen to produce electricity directly.

[1] The term fuel is used in the general context of a flammable, easily ignitable substance.

From balloons to rockets...

In the 16th century the Swiss philosopher and medical doctor Paracelse, also an alchemist and a physicist, wondered if the gas emitted when vitriol and iron react chemically is identical to the air we breathe. Bayle was to isolate this gas in the next century.

Henry Cavendish, British physicist and chemist, repeated the experiments of Paracelse using various metals. In 1766, he collected a large quantity of gas in pork bladders, and showed that this gas, "flammable air", burns in the atmosphere, producing water in the process.
Antoine de Lavoisier, assisted by Laplace and Meunier, was able to explain Cavendish's result after synthesizing water in front of Sir Charles Bogden, secretary of the Royal Society, on June 24, 1783. The experiments led to a letter to the French Academy of Science which concluded that "water is not a fundamental substance, and that it is composed, pound for pound, of flammable air and vital air". This flammable air was henceforth named "hydrogen", which means "that which produces water".
In April of the following year, Lavoisier and Meunier presented to the French Academy of Science an industrial process for the large-scale production of hydrogen by pouring water on red hot iron.
Finally, in 1804, Louis Joseph Gay Lussac and Alexander von Humboldt both showed that water is composed of one volume of oxygen for two volumes of hydrogen.
Cavendish showed that hydrogen has a low density, a fact rediscovered by the physicist and chemist Charles through experiments with soap bubbles.
Many other discoveries followed: in 1839 William Grove demonstrated the principle of the fuel cell. At the end of the 19th century, Pictet and Cailletet independently succeeded in liquefying oxygen. Professor Wroblewski from the University of Krakow was the first to liquefy hydrogen. On May 12, 1898, James Dewar was able, to store liquid hydrogen in a stable, static bath. Finally, Fritz Haber (Nobel prize in 1918) and Carl Bosch (Nobel prize in 1931) invented a process to synthesize ammonia using hydrogen and nitrogen; a process perfected by Georges Claude, the founder of the company Air Liquide.

At the beginning of the 19th century, energy uses of hydrogen first appeared with the introduction of town gas, a gas containing 40 to 60% hydrogen, to light Paris in 1803, based on a process invented by Lebon. Town gas was used extensively until the middle of the 20th century for heating.
Hydrogen began to be used as a fuel for rocket engines in the Atlas-Centaur rocket in 1962, increasing the efficiency of the engines by 50%, sixty years or so after Konstantin Tsiolkovsky first suggested it. The advantage of hydrogen over other rocket fuels is clearly demonstrated by the superiority of the Saturn V rocket over the Soviet N1 rocket. The Saturn V rocket used hydrogen in two of its stages, allowing man to reach the moon in July of 1969, whereas the Soviet N1 rocket relied on conventional fuels. In Europe, work on the use of hydrogen in space exploration began in 1960, culminating in the first successful launch of the rocket Ariane in 1979. Today, all main satellite launchers depend on at least one cryogenic stage using hydrogen.
Nowadays, hydrogen is not used on a large scale as an energy vector, but as an important chemical in the process industry, to produce ammonia, fertilizers and methanol. It is also used extensively for oil refining, in metallurgical processes, as well as in the glass (optical fibres, float glass, etc.) and electronics industries.

...the long history of hydrogen

Henry Cavendish

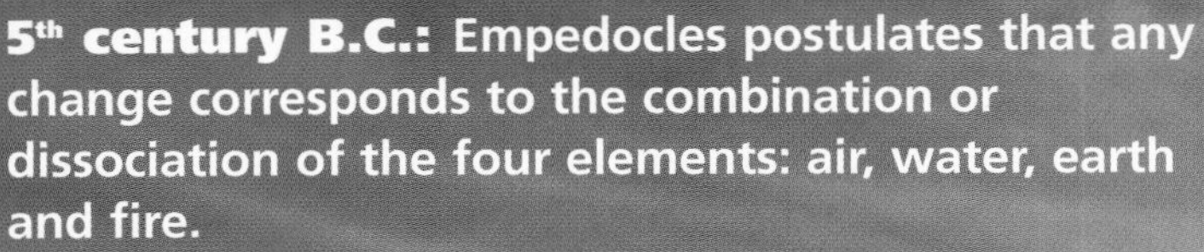

5th century B.C.: Empedocles postulates that any change corresponds to the combination or dissociation of the four elements: air, water, earth and fire.
In the 16th and 17th century: Paracelse and Bayle discover a peculiar gas that would one day be named hydrogen.
1766: Cavendish shows that this gas burns in air, thereby producing water.
1783: Lavoisier synthesizes water from hydrogen and oxygen.

Antoine de Lavoisier

1787: Cavendish followed by Charles shows that hydrogen has a low density.
1789: Van Trootswyck achieves the electrolysis of water.
1803: Lighting by town gas (mostly hydrogen) is introduced in Paris
1839: Grove discovers the principle of the fuel cell.
1909: Haber synthesizes ammonia ; hydrogen then becomes a basic chemical for the process industry.
During the 20th century: hydrogen is used in the petrochemical industry.
1960: hydrogen becomes the fuel of choice for rocket engines used in space exploration.

William Grove

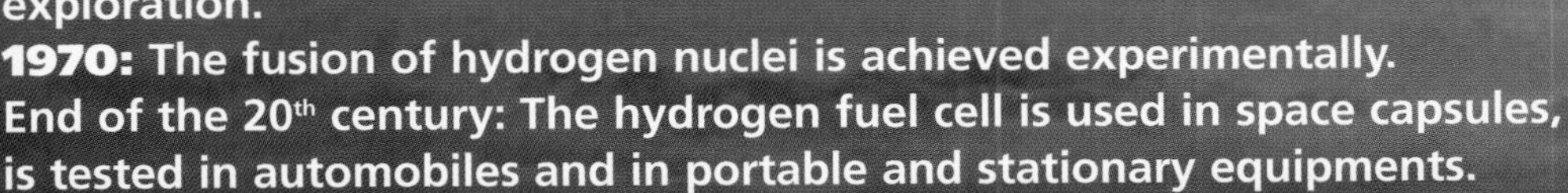

1970: The fusion of hydrogen nuclei is achieved experimentally.
End of the 20th century: The hydrogen fuel cell is used in space capsules, is tested in automobiles and in portable and stationary equipments.

The experiments of Charles inspired the Montgolfier brothers, paper manufacturers in Annonay near Lyon, who were fascinated by the possibility of flying. They made several attempts to use "flammable air" to inflate small balloons of paper and silk, but found that they could only ascend for brief periods of times because the gas could not be contained. They eventually succeeded using warm air. Charles, however, persevered and made a successful flight in front of 300 000 spectators on August 27, 1783 in a balloon filled with hydrogen. This led, in June 1794, to the first military aerial observations by the balloon "l'Entreprenant", which reached an altitude of 300 meters, allowing the armies of the French revolution to thwart the manoeuvres of the Austrian army and to win the battle of Fleurus.

The internal combustion engine

The combustion process in an internal combustion engine (ICE) occurs inside a chamber. External combustion engines were used in the past to power steam boats and the old steam powered trains.
Almost all cars use what is called a four-stroke combustion cycle gasoline engine.
The four strokes are:
_ Intake stroke;
_ Compression stroke;
_ Combustion stroke;
_ Exhaust stroke.

In the intake stroke, the piston starts at the top. The intake valve opens and the piston moves down to let the engine take in a cylinder full of air and gasoline.
The piston then moves back up to compress this fuel/air mixture (compression stroke).
When the piston reaches the top of its stroke, the spark plug fires and ignites the mixture. The gasoline charge in the cylinder explodes, driving the piston down. This is the combustion stroke, which drives the car.
Once the piston hits the bottom of its stroke, the exhaust valve opens; exhaust gases leave the cylinder to go out of the tail pipe (exhaust stroke).
The engine is now ready for the next cycle, and takes in another charge of air/gasoline mixture.
In an engine, the linear motion of the piston is converted into rotational motion by the crankshaft.
The engine described above has one cylinder but most car engines have more (four, six and eight cylinders is quite common), arranged in different configurations: inline, V or flat.
The combustion chamber is the place where compression and combustion take place. As the piston moves up and down, the size of the combustion chamber changes. The difference between the maximum and minimum volumes is called the displacement and is measured in litres or cubic centimeters. If each cylinder of a 4 cylinder engine displaces half a litre, then the entire engine is termed a "2 litre engine". Similarly, a 6 cylinder engine arranged in a V configuration is called a "3 litres V-6". Generally, the displacement determines how much power an engine can produce. A greater displacement is obtained by either increasing the number of cylinders or by making the combustion chamber larger.

Hydrogen combustion engine

In order to use hydrogen in a gasoline engine, some adjustments are required. The most important is to use gaseous fuel injectors for good control of the fuel supply and to produce the lowest emissions and the highest power. The efficiency of the hydrogen engine can be increased more than 4% by using a compression ratio of 14.5/1 because hydrogen has a high octane rating.
Hydrogen ignites very easily. It is not advisable to use platinum spark plugs because they act as catalysts and cause pre-ignition. Iridium plugs are preferred. One has to be careful not to run hydrogen in a "dirty" engine because carbon deposits can act as ignition sources and cause pre-ignition.
Hydrogen burns about six times faster than gasoline. Therefore, the tuning of a hydrogen fuelled engine is more critical because of the high burn rate of the fuel. Nitrous oxide (NOx) emissions of less than 1 ppm from a hydrogen engine can be achieved by proper engine design and engine controls.

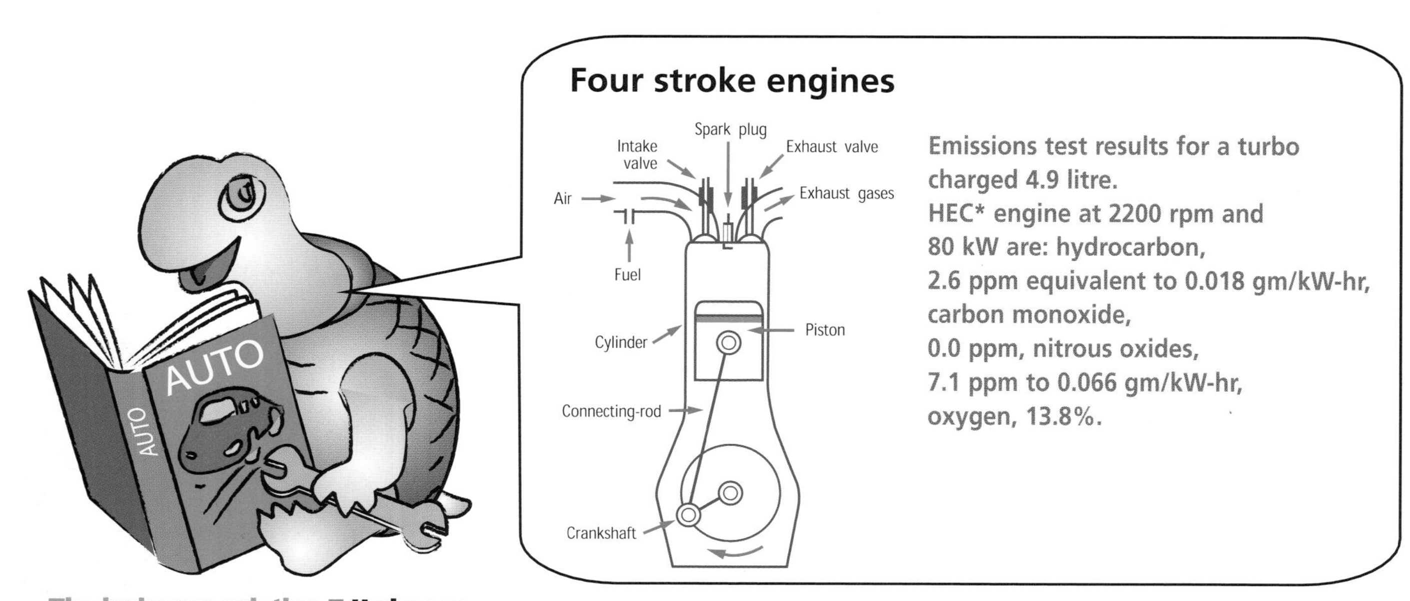

...fuelled by hydrogen

The combustion process in an internal combustion engine (ICE) occurs inside a cylinder. Almost all car engines are four-stroke combustion cycle engines. Hydrogen can be used to fuel an internal combustion engine by adjusting the compression ratio and using a gaseous injector.

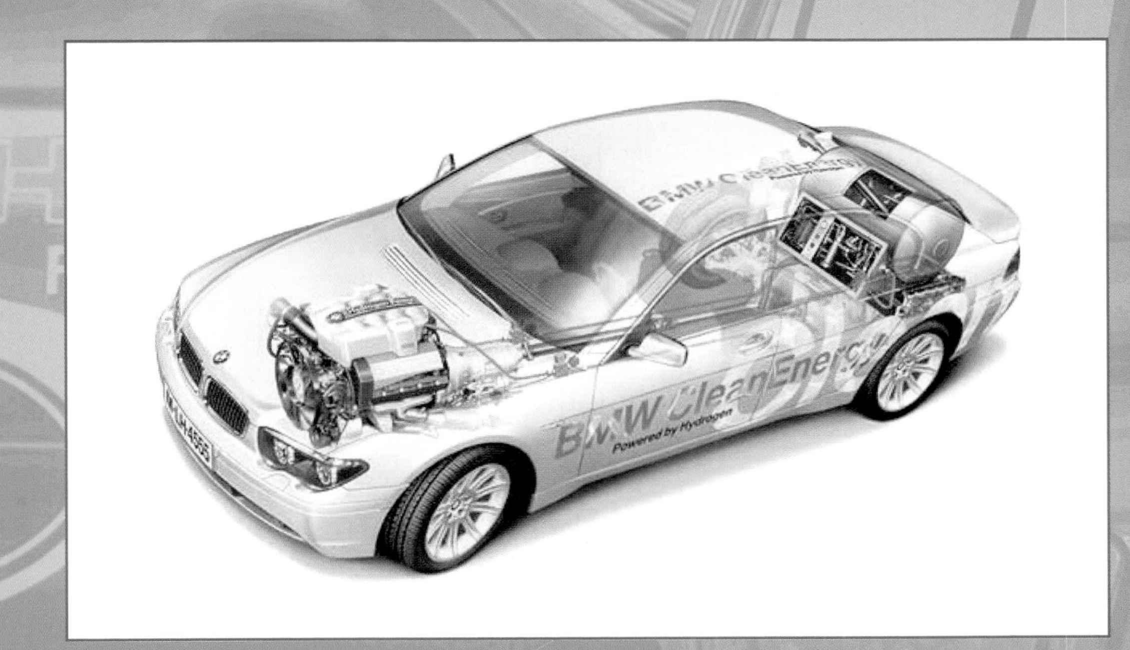

Hydrogen to electricity...

Hydrogen, through an electrochemical reaction with oxygen produces electricity, water and heat[1]:

$$H_2 + 1/2\ O_2 \rightarrow H_2O$$

The electricity is generated from the electrons emitted during the formation of positive hydrogen ions and negative oxygen ions which combine to form water. This electrochemical reaction is the reverse of electrolysis, in which an electrical current running through water produces oxygen and hydrogen.

A device used to produce electricity in this manner is called a fuel cell. The fuel cell consists of an anode and a cathode separated by a conducting electrolyte (acidic or basic). Hydrogen is fed to the fuel cell at the anode, while oxygen flows into the cell at the cathode. Each hydrogen molecule arriving at the anode is dissociated into two protons[2] and two electrons through the reaction:

$$H_2 \rightarrow 2\ H^+ + 2\ e^-$$

The electrons flow toward the cathode through an external electric circuit. This electric current constitutes the electricity produced by the fuel cell, which can be used in external applications. The protons flow through the electrolyte towards the cathode, where they recombine with the oxygen ions produced at this electrode, creating a molecule of water according to the reaction:

$$1/2\ O_2 + 2\ H^+ + 2\ e^- \rightarrow H_2O.$$

The dissociation reactions of hydrogen and oxygen molecules are initiated by a catalyst deposited on each electrode.

A fuel cell unit generates a voltage of 0.5 to 1.0V for a current of 0.5 A/cm^2. As in conventional electric cells, the area of the elements and their stacking determines the active surface which delivers the desired current and voltage. The minimum specific volume of these stacks is currently of 1 litre per 2 kW of power output (with a weight of 1 kg per 1.4 kW). The efficiency of the conversion of chemical energy into electrical energy by a fuel cell is typically about 50%. The overall efficiency (making use of both the electricity and the heat produced) can reach 80%. Using the heat produced by fuel cells for cogeneration is therefore of great interest.

Discovered in 1839 by William Grove, the fuel cell was essentially forgotten until 1960, when it was perfected and used by NASA to supply electricity to the first Gemini space capsules. But it is only since the first oil crisis in the 70s that the fuel cell began to be considered as an energy conversion device that could be used to power vehicles or to produce electricity for stationary applications.

The most common fuel cell is the polymer electrolyte membrane (PEM) fuel cell, also called the proton exchange membrane fuel cell. The PEM fuel cell is composed of an organic membrane that acts as electrolyte and separates the electrodes. This membrane, which has a thickness of a few hundred microns, is made of a carbonated perfluorated polymer permeable to protons containing sulfonic groups. The most common membrane used in PEM fuel cells is Nafion, a trademark of Dupont de Nemours. The catalysts used on the electrodes must resist the strong acidity of the membrane. Today, only platinum can meet these requirements, contributing, with the manufacturing cost of the membranes, to the high cost of fuel cells, which has so far prevented their large scale commercialization.

Direct methanol fuel cells (DMFC) are a variant of PEM fuel cells that operate directly with methanol. Hydrogen is extracted from methanol by a ruthenium catalyst at the anode. Methanol is a liquid, which is easier to store than hydrogen. DMFC could thus be used to power portable applications such as laptops or cell phones.

The PEMFC is considered by most automakers to be the prime candidate to convert hydrogen into electricity for light electric vehicles powered by hydrogen. Therefore, significant efforts are being directed toward increasing the performance of PEMFC and reducing their cost. Several automotive applications of PEMFC have been achieved. Small fleets of 20 to 60 hybrid vehicles have been made available for rental since 2003. Many other PEMFC applications have been realized, with power ratings ranging from the watt to the megawatt.

[1]Due to combustion
[2]The two H^+ ions.

...with the fuel cell!

The electrochemical combustion of hydrogen in a fuel cell produces water, heat and electricity. Fuel cells are composed of an electrolyte which separates the anode (negative pole) fed by hydrogen from the cathode (positive pole) supplied with oxygen.

Fuel cells are essentially noiseless and can produce electricity without emitting pollutants or greenhouse gases. They can be used in a wide range of applications. Fuel cells are one of the major applications behind the great interest in hydrogen.

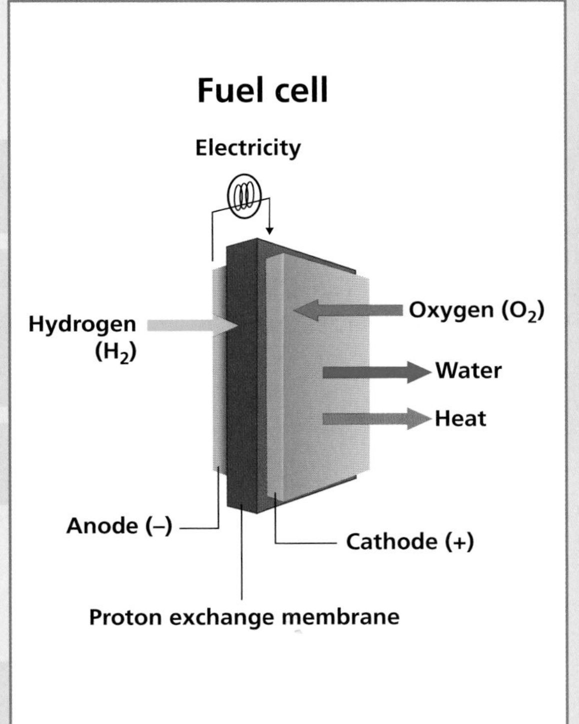

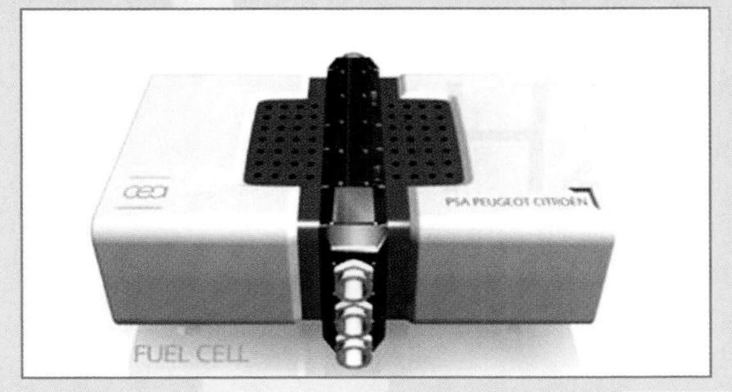

Different types of fuel cells...

The Alkaline Fuel Cell (AFC) is a liquid electrolyte fuel cell. The liquid electrolyte is typically a 8 to 10 mol/litre potash (potassium hydroxide) solution. The AFC typically operates at a temperature of 80 to 90 °C. It can operate at higher temperatures when pressurized (200 to 230 °C for the Bacon fuel cell) or when using a more concentrated electrolyte (Apollo fuel cell). The potash of the electrolyte can flow or be set in a membrane. The hydroxide ions can form carbonates by combining with CO_2, resulting in a lower conductivity of the electrolyte. Thus CO_2 cannot be present in the oxygen or the hydrogen feeding the fuel cell. This type of fuel cell was essentially developed for the aerospace industry and has been used successfully during the manned missions of NASA. Its simple design results in lower manufacturing costs. Its main drawback is the high level of purity required for the feedstock gases.

The Phosphoric Acid Fuel Cell uses phosphoric acid embedded in a matrix as its electrolyte. This acid does not react with CO_2. The phosphoric acid fuel cell can thus use non purified gases as feedstock. Its low vapour pressure allows operation at high temperature, resulting in higher efficiency. However, the electrolyte solidifies at 42 °C. This requires that it be maintained above this temperature in order to avoid solidification, which would increase its volume and strain the overall structure of the fuel cell. The electrodes are made of thin carbon films covered with platinum to resist the chemical reactivity of the electrolyte. This fuel cell was successfully developed by ONSI Corp, which has had some success in commercializing stationary units with power ratings ranging from 25 to 200 kW (with more than 2 million hours of operation accumulated).

The Molten Carbonate Fuel Cell is a fuel cell whose electrolyte is a eutectic mixture of lithium carbonate and potassium. Its operating temperature is 650 °C, at which point the reaction kinetics are favorable to hydrogen oxidation and oxygen reduction. The anode is made of porous nickel doped with a few percent chrome. The cathode is also made of nickel, covered with lithium doped with nickel oxide.
There exist several molten carbonate fuel cell prototypes with power ratings ranging from a few tens of kW to 2 MW, built and tested for many years by the Daimler subsidiary MTU, FuelCell Energy (USA) and MCFC Research Association (Japan). Its drawbacks are the high temperature of the liquid electrolyte and the risks of corrosion. Its advantage is the possibility of using coge-neration to increase the overall efficiency of the fuel cell.

The Solid Oxide Fuel Cell (SOFC) uses a solid oxide electrolyte which, in order to be a good ionic conductor, must be kept at a temperature between 800 and 1000 °C. The reduction of oxygen at the cathode produces ions which migrate through the solid electrolyte until they reach the anode, where they combine with the hydrogen ions at this electrode, thereby producing water. As for other fuel cells, the electron flow released goes through an external electric circuit, which corresponds to the electrical current provided by the fuel cell. The electrolyte generally used is zirconium doped with yttrium. The addition of 8 to 10% yttrium ensures sufficient ionic conductibility of the oxygen ions. The anode is commonly made of nickel. The cathode, must be chemically stable in an oxidizing medium, have good electronic conductivity and a sufficient catalytic activity to dissociate oxygen. It is generally made of $(La,Sr)MnO_3$ (lanthanum manganite doped with strontium). This fuel cell generally uses natural gas as a feedstock, converted into hydrogen at the anode by high temperature reforming.
The Solid Oxide Fuel Cell has generated a great deal of interest because of its high overall efficiency (exceeding 80% with cogeneration). This level of efficiency is due to the fact that it releases heat at high temperatures -about 800 °C- during its operation, which can be harnessed through combined cycles (a turbine, followed by thermal treatment of the by-products at temperatures of 300 to 400 °C). The development of SOFC technology has been the object of a great deal of effort in the United States, Japan and Europe. Cylindrical solid oxide fuel cells developed by Siemens-Westinghouse have been in use for more than 70,000 hours. Prototypes of 100 and 200 kW have been in successful operation since 1998. A planar technology, more complex but potentially more compact, is being developed in Japan and in Europe.

...for diverse applications

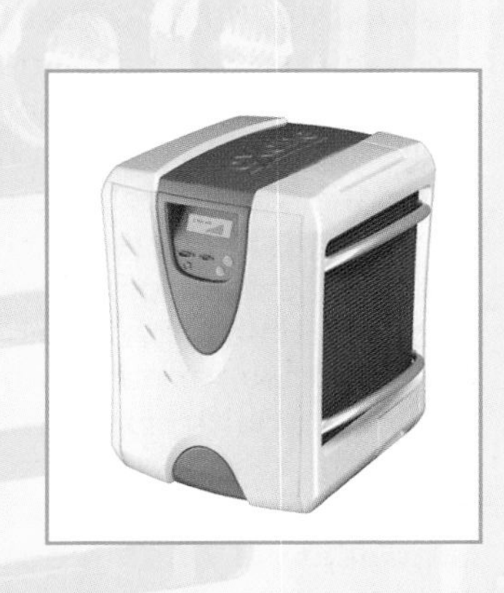

Type of fuel cell	Electrolyte	Operating temperature	Utilization
Alkaline Fuel Cell (AFC)	Potash (Liquid)	80 °C	Space, transportation Range : 1 - 100 kW
Proton exchange Membrane Fuel Cell (PEMFC and DMFC)	Polymer (Solid)	80 °C	Portable, transportation, stationary Range : 1W - 1 MW
Phosphoric acid Fuel Cell (PAFC)	Phosphoric Acid (Liquid)	200 °C	Stationary, transportation Range : 200 kW - 10 MW
Molten carbonate Fuel Cell (MCFC)	Molten salts (Liquid)	650 °C	Stationary Range : 500 kW - 10 MW
Solid oxide Fuel Cells (SOFC)	Ceramic (Solid)	700 to 1,000 °C	Stationary, transportation Range : 1 kW - 10 MW

Hydrogen on the road...

Our land transportation system currently relies on gasoline. The transportation sector is responsible for about 30% of the world's anthropogenic production of CO_2, when emissions due to oil refining are included. This number is expected to grow as developing economies emerge. Electric vehicles, whose range is limited by their capacity and their charge time, are expected to be of practical use only for very specific applications. The use of hydrogen powered engines (either thermal engines or fuel cells) in vehicles such as cars, buses, trucks and two-wheelers could lower the contribution of land transportation to anthropogenic CO_2 emissions.

The classic engine

In a classic gasoline or diesel powered piston engine, the working fluid used to convert thermal energy into mechanical work is a mixture of the gases produced by the combustion of the fuel with the oxygen in air, and of inert nitrogen which constitutes 79% of the air.
When the hot and compressed gases expand, they transfer their mechanical energy to the pistons. Using hydrogen instead of gasoline does not change this operating principle. However, certain modifications are required to adapt the internal combustion engine to hydrogen, particularly with respect to the direct injection of the fuel in the cylinders. This is because:
- During the injection, hydrogen occupies a relatively large volume, which reduces the quantity of working fluid for each cycle, leading to a power output 20 to 25% smaller;
- Hydrogen is very sensitive to auto ignition and backfires.

The high temperatures generated by the combustion of hydrogen lead to an efficient engine, but can produce some nitrogen oxides (NOx), in addition to water vapour. However, lean burning of hydrogen could practically eliminate the production of nitrogen oxides.
Because the internal combustion engine has been in use for over a century, there is a considerable body of experience in designing and producing them. Thus, hydrogen internal combustion engines could already be mass produced within the current industrial infrastructure. In addition, these engines could also be made to work in parallel with gasoline, leading to the construction of bi-fuel engines, which could be developed before the establishment of a hydrogen refueling infrastructure. Several hydrogen-fuelled and bi-fuel prototypes have been produced in the United States, Europe and Japan.

The fuel cell electrical engine

An engine relying on a fuel cell requires several subsystems. First, a compressor is needed to provide air to the fuel cell. A radiator is also required to evacuate the heat from the fuel cell and the electric system. Finally, a traction inverter module is needed to convert the electricity for use by the electrical motor transaxle, which transforms electricity into mechanical energy.
In addition to the inherent efficiency of the fuel cells (above 50%), an electric vehicle does not consume energy when stopping, and some energy can be recovered during braking. On the other hand, the auxiliary systems require a non negligible amount of energy, reducing the overall efficiency to 40%.
However promising the fuel cell is for the transportation sector, it is still too expensive, and has only been implemented so far in demonstration prototypes using mostly PEM fuel cells. Ballard Power Systems pioneered the use of fuel cells in the transportation sector by building 6 buses equipped with a 200 kW fuel cell, which were tested for two years in Vancouver and Chicago. American, European and Japanese companies have since designed and tested several hydrogen fuel cell prototypes (cars, buses, utility vehicles). Prototype electric two-wheelers based on hydrogen powered fuel cells have even been built, such as scooters, motorcycles, etc.

The hybrid architecture

The idea, famously applied to the Prius by Toyota, is to use a hybrid power system combining an internal combustion engine and an electric engine. When the power demand is low, the ICE recharges the batteries. Otherwise, the batteries produce part or all of the power required by the car, lowering harmful emissions. The ICE thus operates only in a regime where it is most efficient, leading to a smaller ICE required to power the vehicle. This could be applied to a hydrogen fuelled vehicle, which would emit no CO_2. A hybrid concept could also be adapted to a vehicle using a fuel cell instead of an ICE, which could reduce the size and the cost of the fuel cell required.

On rails without overhead lines

Some manufacturers such as IRIBUS are considering fuel cell powered tramways that would no longer require overhead lines. This idea could be generalized to electric trains, which could then run on rails currently used by Diesel trains.

...cars, tramways and the hydrogen train

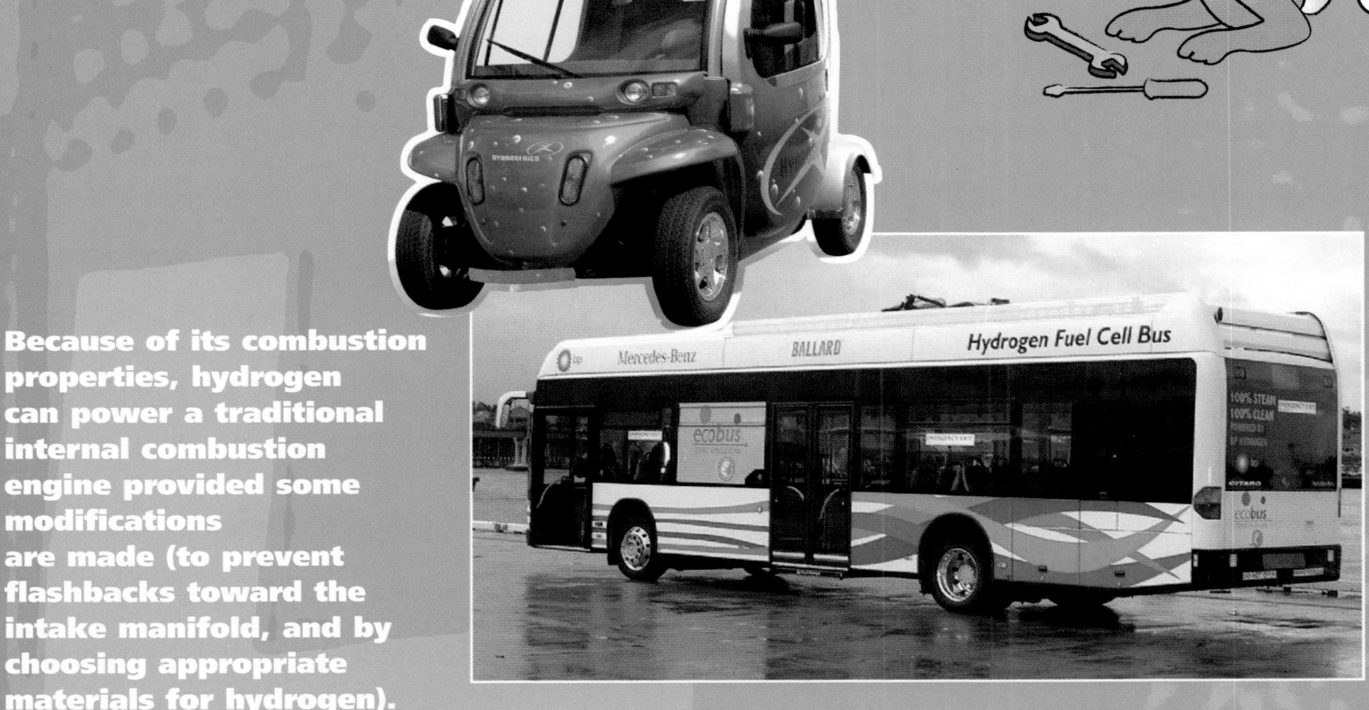

Because of its combustion properties, hydrogen can power a traditional internal combustion engine provided some modifications are made (to prevent flashbacks toward the intake manifold, and by choosing appropriate materials for hydrogen).
Such engines do not emit carbon dioxide, but only water vapour and a small amount of nitrogen oxides as by-products of combustion.
Hydrogen-based electrical vehicles are powered by a fuel cell. A fuel cell based engine is essentially noiseless, emits no pollutants, has a longer autonomy and a short refueling time compared to charging battery powered electric vehicles.
Virtually any vehicle can use fuel cell engines: from the bus to the scooter, including trucks, and utility vehicles.
Tramways and trains using a fuel cell based electric engine no longer require overhead lines.

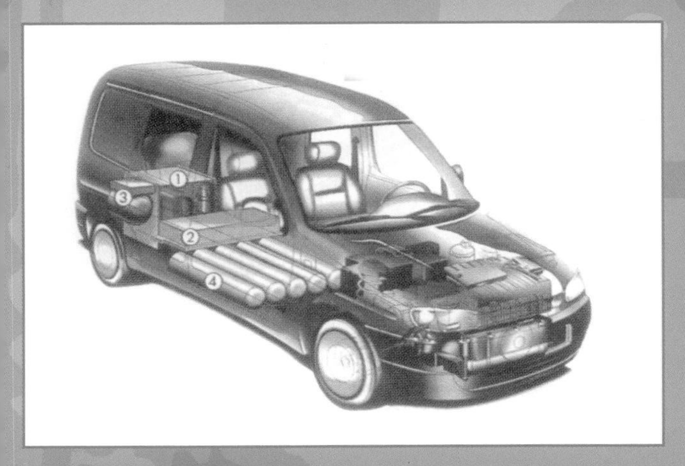

1, PEM fuel cell; 2,3 accessories;
4, hydrogen reservoirs.

On the sea and underwater...

Clean, quiet boats, barges and ships

Boats can navigate on water with few limits on their bulk or weight. It is readily conceivable that hydrogen could be used to power boats using either an electric engine powered by a fuel cell or a thermal engine. Fuel cells are well suited for small boats. Fuel cell powered boat prototypes can be found on Swiss lakes. Thermal engines could be used to power barges or ships.
Access regulations to certain bodies of water, such as the Baltic Sea, are being tightened in order to limit the pollution due to the many ships powered by fossil fuels that navigate these waters. In addition, classic propulsion systems are noisy and generate vibrations, which can be detrimental to the working environment of the crew or to the comfort of passengers. They can also make oil exploration and underwater detection more difficult. Finally, noise and pollution restrict the access to certain protected zones such as natural preserves or tourist sites.
In addition, hydrogen can be produced locally. Iceland is considering transforming its important fishing fleet to work with hydrogen produced from geothermal and hydraulic sources. This could save 275,000 tons of oil per year and lower CO_2 emissions by 720,000 tons per year, or half a ton of oil per ton of fish pulled out of the sea.

Submarines with increased autonomy

Most submarines have been used by the military. They are expected to operate discretely and remain as long as possible underwater. This has led to the development of nuclear submarines that can operate for months without surfacing. Submarines powered by standard propulsion systems rely on an electric engine powered by batteries which are recharged when the submarine operates on the surface, using diesel engines. Recharging must occur frequently because of the low autonomy of the batteries. Such submarines can only remain submerged for a few days. This operation is risky because the submarine remains relatively visible and can be detected from the noise of its engine, when even it remains submerged using a snorkel[1].
With hydrogen fuel cells, a submarine could remain submerged for a month instead of days. They have been used by the German navy since 2004 in the operational prototype U-31. This submarine can operate 3000 km under water at low speed. The ship, 57 meters long and manned by a crew of 27, uses a combination of PEM fuel cells and batteries to achieve a power rating of 300 kW. The feedstocks of the fuel cells must be stored on-board. Oxygen is kept as a cryogenic liquid, whereas hydrogen is in a metal hydride (FeTi). Both are stored in cylinders located between the hull of the ship and its superstructure. The metal hydride reservoir weighs 160 tons but holds only 1800 kg of hydrogen. The weight of the metal hydride storage system is not a problem as it is also used as ballast.
Many countries such as Greece, Italy, Portugal and South Korea are currently considering purchasing such systems, which could be installed as upgrades to their existing fleet.
Besides military applications, hydrogen powered submarines are also being used for undersea exploration because of their autonomy. The Japan Marine Science and Technology Center launched the Urashima 600 in December 2000, which achieved a run of 220 km, 800 meters below the surface in June of 2004.

[1]A tube that allows the operation of a diesel engine underwater, by evacuating burnt gases and taking in fresh air.

...hydrogen for navigation

A hydrogen boat can, like a vehicle, be powered by a thermal engine or a fuel cell electric engine. Because there are few limitations on volume or size for boats, storing hydrogen onboard is not as problematic as for vehicles.
A hydrogen submarine can also operate using an electric engine powered by a fuel cell, which allows it to move essentially without noise or production of toxic combustion by-products.
On-board, hydrogen would also make possible the decentralized production of electricity for auxiliary applications to enhance safety on large ships or to power the stem propeller.

A 100-ton fishing boat powered by a 500 kW engine would require 1000 kg of hydrogen for a 4 to 5-day outing. Such a quantity of hydrogen could be stored on-board either as a liquid or in metal hydrides (discussed in the next section). Hydrogen could also be used to provide the electricity required by the ship's operation, to provide emergency power and, in large ships, to power the stem propeller used for manoeuvres, increasing safety by eliminating extensive networks of cable.

In the air...

Airplanes are responsible for 12% of the CO_2 emissions of the transportation sector, and for 3 to 4% of the overall anthropogenic emissions of CO_2. Their contribution is likely to increase as air travel is expected to grow by about 5% per year for decades to come, due to the emergence of developing economies.

Using liquid hydrogen instead of kerosene

Why not consider, then, a hydrogen fuelled airplane in the medium to long term? For a given amount of energy, hydrogen is 2.8 times lighter than kerosene but 4 times bulkier even in its liquid form. Its eventual deployment hinges on the possibility of designing planes that could store a lighter, bulkier cryogenic fuel.

Discrete and more efficient engines...

The turbojet engine, based on the gas turbines used in aeronautics, is an example of a thermal engine that can be adapted to hydrogen. Because hydrogen has higher energy content than kerosene, a hydrogen powered plane would carry less fuel mass. It would be lighter at take-off, and the required power rating of the engines could be lowered, thus producing less noise. In addition, by taking advantage of the low temperature of hydrogen, the turbojet engines could be cooled, increasing their efficiency by 5 to 10%. Due to its high diffusivity, hydrogen can mix with air more quickly. The combustion chambers could then be smaller and the weight of the engines reduced. Finally, as a clean fuel, hydrogen could increase the lifetime of turbojets by 25% and lower their maintenance rate.

But a more complex fuelling

On the other hand, feeding hydrogen to the engine is a complex process. Low temperature liquid hydrogen must be brought to the combustion chamber of the engines, where it must be injected in gaseous form at ambient temperature. The cryogenic fuel thus must go through a series of heat exchangers, cooling the turbine and other systems. It is then warmed up (by the burned gas from the engine) and compressed to ensure sufficient outflow. In order to achieve this, hydrogen, still in the liquid state, is compressed to 3-5 MPa using centrifugal pumps located in front of the heat exchangers. Because of the low density of hydrogen, the pumps must rotate at high frequency (several tens of thousands of turns per minute).

And bulkier reservoirs

The liquid hydrogen reservoirs must be well insulated and designed in such a way as to minimize the surface to volume ratio. Thus, the ideal configuration would be spherical. However, a cylindrical shape is also manageable. The reservoirs could be located inside the fuselage, at either extremity of the plane (with a quasi-spherical shape), or on the top of the cabin (with a cylindrical shape). The liquid hydrogen reservoirs must be designed to prevent air from coming into contact with its content because it would solidify. The reservoirs must always be pressurized with hydrogen, at a pressure slightly above the atmospheric pressure at sea level.

The airplane of the future?

The hydrogen airplane would have a surprising shape, bulkier but with smaller wings because of its lower weight. In order to introduce the hydrogen airplane in a very competitive market, the concerns of the public and cost issues will have to be addressed. The logistics of airports would have to be reviewed.
The hydrogen airplane would offer the possibility of increased storage capacity and low environmental impact.

...The hydrogen plane

Another possible application of hydrogen thermal engines is the turbojet engine (a gas turbine adapted for use in the aerospace industry). The design of such systems for use with hydrogen must take into account its low density, its explosive nature, and the fact that it must be stored as a cryogenic liquid for an acceptable operating range.

Hydrogen on the ground in the airport

The Munich Airport in Germany has used hydrogen buses for many years. Certain buses rely on internal combustion engines and others on fuel cells. A hydrogen fuelling station is even available on site. A Canadian project at the Montreal airport has been proposed. They intend to use three fuel cell people movers, 10 service vehicles (tugs) and three buses using internal combustion engines powered by hydrogen.

In space...

Launching a satellite

Putting a satellite in orbit requires sending several tons of equipment several hundreds of kilometers above the Earth in less than 30 minutes at a velocity of 30,000 to 36,000 km/hour, while overcoming friction forces and the weight of the satellite. This requires high performance rockets with several (generally three) stages, successively activated and ejected.

Propulsion

The last stage of the Ariane 5 rocket illustrates well the basic principles of propulsion with a rocket engine. It consists of a reservoir containing both propellants - fuel and oxidant- used by the engine, which also has a structural role. The payload is found at the front, whereas the rocket engines are located at the back. The thrust is produced by the high velocity ejection of the combustion products, which are accelerated by going through a nozzle. When drag forces are no longer exerted on the launcher, it behaves as an isolated system. At every instant, the momentum[1] lost by the launcher through the gas ejected at the back is compensated by an increase in its velocity in the opposite direction. This leads to the following:

- The thrust produced by a rocket engine is the product of the velocity of the ejected gases and the flow rate of the propellants consumed by the combustion process. In other words, for a given value of thrust, the faster the ejection velocity, the lower the consumption rate of the fuel. Since 90% of the weight of the system is due to the fuel and the oxidant, a greater ejection velocity is preferred.
- The velocity gained from a given stage of the rocket is proportional to the ejection velocity of the combustion products.

Hydrogen's golden path

With classical propellants such as kerosene and liquid oxygen, the ejection velocity of the combustion by-products can reach at most 3,300 m/sec. Liquid hydrogen and oxygen can produce significantly greater ejection velocities (4,500 m/sec). The gain is such that replacing a single classical stage using kerosene by a cryogenic[2] stage based on hydrogen can, under certain conditions, result in increasing the mass of the payload by 50%!
The interest of using hydrogen is not restricted to the last stage of a launcher. The European Ariane 5 rocket, initially designed with a classical stage located on top of a large main cryogenic stage, now possesses two cryogenic stages that use hydrogen. The number of satellite launchers with two cryogenic stages is increasing worldwide.

Performance conditions

**The ejection velocity is greater when the expansion rate in the nozzle is high. It is proportional to the square root of the combustion temperature and inversely proportional to the molecular mass. It is the low molecular mass and high combustion temperature of hydrogen which justifies its use as a rocket fuel.
Because of its low density, however, liquid hydrogen requires large and relatively heavy reservoirs. In addition, the pumps, through which run large volumes of fluid, are subjected to a considerable amount of power and must also rotate very fast to provide a sufficient centrifugal effect. In addition, the ball bearings cannot be lubricated. The hydrogen turbo-pump of Vulcain 2, the engine of the Ariane 5 rocket, runs at 35,000 turns per minute, compared to 60,000 for the top stage, and can produce a power of 14 Megawatts while weighing only 250 kg.**

[1]Momentum is the product of the velocity and the mass of an object.
[2]Hydrogen and oxygen can only remain liquid below 20 and 91 Kelvin respectively at atmospheric pressure, which explains why the stage is termed cryogenic.

...Hydrogen as the fuel of choice for rockets

The hydrogen-fuelled rocket engine used in space exploration is perhaps the best known use of hydrogen as an energy carrier.

The thrust is generated by the expansion of high temperature, high pressure gases in a nozzle from a combustion chamber continuously fed by liquid hydrogen and liquid oxygen reservoirs.

Electricity from hydrogen ...

Power generation for portable applications

The usefulness of cell phones (operating at about 100 mW) and of laptop computers (30 W) is currently limited by the autonomy of their lithium-ion batteries, which allow operation for a few days in the case of cell phones and for about three hours for laptop computers. In order to increase the autonomy of portable applications, intense research and development activities in Canada, the United States, Japan, France and Korea have been directed towards the development of a micro fuel cell coupled with a Li-ion battery as a power source. Autonomy would then be only limited by the supply of hydrogen. The refill could be done in a few seconds by changing a cartridge, as we do for a pen or a lighter. Each refill would yield autonomies 3 to 5 times greater than the current Li-ion battery. The fuel cells considered for this application are either PEMFC or DMFC because they operate under ambient conditions and offer a solid-state solution easily integrated in portable systems.

Emergency Power Generators

Emergency Power Generators (EPG) are basically fuel cells with power ratings of the order of a few hundred Watts to a few kilowatts used to provide power in the absence of a grid. Some manufacturers already commercialize PEM and DM fuel cells which can replace classic EPGs powered by thermal engines fuelled by gasoline. EPGs would power hospitals and computer centres in case of power outage, or equipment in isolated areas. They require little maintenance and are therefore well suited for such uses. However, the hydrogen internal combustion engine (HICE) could also be used as EPG with little or no NOx emissions, provided hydrogen combustion is controlled to a lean burning mode.

Auxiliary Power Units (APU)

Auxiliary Power Units are generators used on-board vehicles to power their electrical equipment. They are autonomous and do not depend on the vehicle's engine. On commercial airplanes, the APU is typically a gas turbine connected to a generator, which can be used even when the reactors are stopped. The possibility of replacing it with a fuel cell is currently being considered by airplane manufacturers. It is, however, in the context of the Apollo lunar exploration program that the first APU aerospace application occurred. The fuel cells could provide enough electrical power without significantly increasing the weight of the vehicle, which could not be done with conventional batteries.

The energy required by electrical equipment in cars is not negligible compared to the energy required for propulsion. This has lead to providing two sources of energy for the two functions. For instance, BMW hydrogen cars, powered by an internal combustion engine, use a fuel cell as the APU.

Hydrogen APUs could be found on the sea, in competitive sailing boats, and on large ships where decentralized production of electricity can eliminate the need for potentially unsafe large electrical wiring.

Residential applications

Fuel cells can be used to supply electricity in isolated areas not connected to a grid, particularly if the hydrogen can be produced locally, using wind turbines. For larger scale applications, the heat from the fuel cells can also be used. Fuel cell installations used as stationary power generator can produce collectively 200 kW to a few MW or from 1 to 5 kW if a single fuel cell stack is used. They can be connected to the grid using an inverter which allows the optimization of their size by including the possibility of buying or selling electricity. The units would preferably use easily available natural gas, which requires a reformer to produce the hydrogen needed by the fuel cells. Solid Oxide Fuel Cells would be ideal for such applications, because they can be fuelled directly with natural gas and do not require a reformer.

...for a greater autonomy

Power generation for portable applications

Fuel cells can be produced in small sizes (with power ratings ranging from one watt to a few hundred watts) and used to supply power to portable applications such as CD or DVD players, laptop computers, digital cameras and cell phones, which could be recharged using a cartridge containing hydrogen, offering much longer autonomy than currently available batteries.

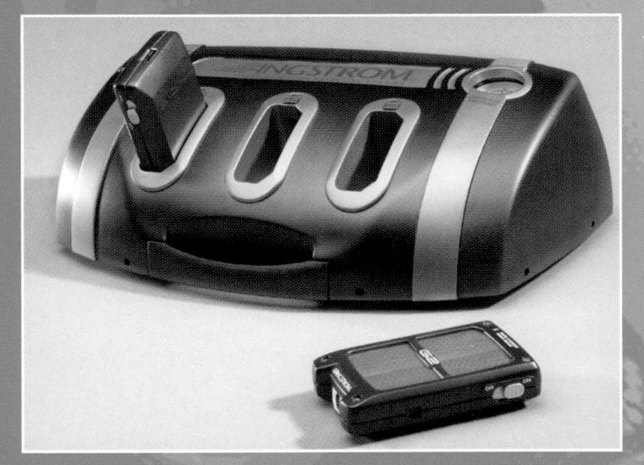

Large scale power generators

Fuel cells can also come in large sizes. They can generate power levels up to the Megawatt, and can be used as alternate power sources to the electric power grid. In particular, they can be used as emergency power generators or to continuously supply electricity to a decentralized network.

HONDA
FCX
BELL
AIR LIQUIDE

Is hydrogen an ideal energy vector?

Hydrogen can be used for heating, lighting and transportation. It can also power electric and electronic circuits. All of this can be done noiselessly without emitting greenhouse gases or pollutants. Hydrogen has many qualities that would make it the ideal fuel. However:

- **Hydrogen does not exist in a free state on Earth and must be produced using an energy source, ideally without harmful emissions.**
- **Hydrogen must be stored in large quantities on-board vehicles, a difficult task considering its low density.**
- **Like fossil fuels, hydrogen must be transported and distributed.**
- **Like any other fuel, hydrogen presents some risks and must be handled with care.**

Using hydrogen on a large scale will require producing it, storing it and distributing it safely and efficiently.

Producing hydrogen...

Although hydrogen is rare in its natural state on Earth, it is widely present in molecular form. In particular, it is found linked to carbon atoms in various chemicals which have a large energy content. Thus, the extraction of hydrogen from these compounds requires little energy input. This way of producing hydrogen is economical, and it is the method most widely used to make hydrogen for the process and petrochemical industries.

Although producing hydrogen from fossil fuels is cost-effective, this approach results in the emission of large quantities of carbon dioxide as a by-product. This makes hydrogen obtained this way less attractive as a fuel because of the reduced environmental benefits, unless the carbon dioxide can somehow be sequestered during production.
It is possible to sequester the CO_2 produced by these methods by geological storage. This would entail storing carbon dioxide in oil or natural gas deposits, either depleted or active (the carbon dioxide could be used to extract oil), or in unused coal veins or deep saline underground water sources.
The sequestration of CO_2 should last at least for the period during which its action as a greenhouse gas is likely to remain critical, which would be a few centuries. However, as a precaution, sequestration should be planned for a duration of several thousands of years. CO_2 sequestration would be acceptable only if it does not adversely affect the environment through leaks into drinkable water systems, in ecosystems and in human habitation areas. Monitoring equipment would also have to be installed for the duration of the sequestration.

The reforming* of natural gas and hydrocarbons

The process can be used for natural gas or light hydrocarbons. Hydrogen is obtained from syngas (synthesis gas), a mixture of hydrogen, carbon monoxide and to a lesser extent, from carbon dioxide, methane and water. Through a chemical process called vapour reforming*, natural gas or light hydrocarbons are transformed into synthesis gas through a chemical reaction involving steam on a nickel catalyst at high temperature (840 to 950 °C) and moderate pressure (about 20 to 30 bar). Typically for methane, the reaction is

$$CH_4 + H_2O \rightarrow CO + 3H_2$$

A second reaction, "water gas shift", converts the CO into carbon dioxide:

$$CO + H_2O \rightarrow CO_2 + H_2$$

This whole process, where hydrogen taken from the carbon and oxygen from water is converted into carbon dioxide, can be summarized in the global reaction:

$$CH_4 + 2H_2O \rightarrow CO_2 + 4H_2$$

The CO_2 and the remaining CO can be separated from the hydrogen in one of two ways. A chemical process can extract 98% of the hydrogen, resulting in a gas 95% to 98% pure. Direct purification by pressure swing adsorption (PSA*) on an adsorbent yields a hydrogen gas 99.9% pure, but can only extract 90 to 95% of the hydrogen in the mixture. A large scale facility (producing 43,000 tons of hydrogen per year) has an energy efficiency of 65%, producing carbon dioxide at a rate of 10 to 11 tons per ton of hydrogen.

Partial oxidation

Another process, partial oxidation, can be used to produce hydrogen from heavy petroleum by-products and coal. Partial oxidation occurs at high temperature (1,200 to 1,500 °C) and high pressure (2 to 9 MPa and more), in the presence of oxygen and a temperature regulator (water vapour), usually without a catalyst. As in vapour reforming, the critical step is the production of the syngas using a reaction such as:

$$C_nH_m + (n/2)\,O_2 \rightarrow n\,CO + (m/2)\,H_2$$

The carbon monoxide is then subjected to a gas shift reaction as discussed above. The energy efficiency is about 55% with carbon dioxide emissions of 15 tons per ton of hydrogen. From an economic standpoint, this method is justified when the cost of the raw material is low (such as petroleum coke for instance).

Thermochemical production of hydrogen from biomasss

Biomass is in its essence a vegetable matter, obtained from the photosynthesis of CO_2 and H_2O, in a process which involves solar energy to produce cellulose, lignocelluloses and lignin, with the chemical composition $C_6H_9O_4$. The first thermochemical treatment in this process is performed in the absence of oxygen (thermolysis at about 500-600 °C), followed by water vapour gasification at 900 °C which results in the production of syngas (CO and H_2). This step is similar to the vapour reforming of hydrocarbons and natural gas discussed earlier. A gas shift reaction then transforms water and carbon monoxide into carbon dioxide and hydrogen:

$$CO + H_2O = CO_2 + H_2$$

This process is CO_2 neutral in the sense that the CO_2 molecules produced by the process do not contribute to the greenhouse effect since they were initially obtained from the atmosphere by plants.

Plasma dissociation

Hydrogen can be produced from natural gas without CO_2 emissions by dissociating natural gas (principally methane) into hydrogen and carbon black by a pulsed electric arc discharge operating at low temperature in cold plasma.

...with greenhouse gas emissions

Hydrogen can be extracted from a wide range of chemical compounds: fossil fuels, water and biomass. From natural gas and other hydrocarbons, hydrogen can be obtained by vapour reforming or partial oxidation. These high temperature chemical reactions occur in the presence of water or oxygen, which exchange the hydrogen atoms linked to carbon atoms for oxygen. Producing hydrogen this way also generates carbon dioxide, which could be sequestered geologically. Biomass can be used to obtain hydrogen through a thermochemical process that produces CO_2 as a by-product, but the reaction is called CO_2 neutral because the CO_2 generated corresponds to the CO_2 that plants had initially absorbed from the atmosphere.

Producing hydrogen without CO_2...

Water is essentially the result of burning hydrogen. Why not reverse the process to produce hydrogen from water? Although more energy is needed to produce hydrogen from water than from hydrocarbons, no CO_2 molecules are emitted by this process.

Hydrogen production by water electrolysis

Electrolysis is the chemical dissociation of molecules through the application of a continuous electrical current. An electrolysis cell is formed of two electrodes, an anode and a cathode. The electrodes are immersed in an electrolyte (an ion conducting medium). The electrolyte can be an acidic, basic or saline solution. In the case of an acidic electrolyte, the hydrogen ion appears at the cathode, the rest of the elements at the anode. In the other cases, the metallic ion appears at the cathode. If the electrodes are chemically stable, secondary reactions occur in their vicinity involving water, resulting in the following global reaction:

$$H_2O \rightarrow H_2 + 1/2\ O_2$$

Electrolysis on an industrial scale is generally performed using an aqueous solution of potassium hydroxide, of varying concentration depending on the operating temperature (25% by wt at 80 °C up to 40% at 160 °C). Potassium is preferred to sodium, since it has a larger electrical conductivity at a given temperature, and offers a better control of chlorate and sulphate impurities. The electrodes are generally bipolar plates used as an anode on one side and as a cathode on the other.

Conduction occurs inside the electrode along its width, thus allowing for a less resistive medium and for a larger current density.
The efficiency of the electrolysis process, starting from the input electric power, ranges from 75% to 85%. As a result, the cost of hydrogen produced using electrolysis is high. The hydrogen produced is, however, very pure. Water electrolysis constitutes an interesting way to store electrical energy produced at low cost (off peak or from dedicated wind turbines). It would not be coherent to use electricity produced from thermal power plants using fossil fuels, because direct and more efficient means exist to produce hydrogen from these sources, while emitting less CO_2!

Thermal dissociation of hydrogen

Direct thermal dissociation of water is possible, but only occurs at very high temperature (2,600 °C). It can be achieved at lower temperatures (800 to 900 °C) through a series of chemical reactions involving water and chemicals such as sulphuric acid and iodine, followed by dissociation processes that result in the production of hydrogen and oxygen. The end products are hydrogen, oxygen and the chemical compounds initially used. A new generation of high temperature nuclear reactors using gas coolants (HTGR) would make it possible to produce hydrogen this way.

Bioproduction of hydrogen by photosynthetic micro-organisms

Certain unicellular green algae (or cyanobacterias) can produce hydrogen from solar energy using water as the donor of electrons and protons, without emitting greenhouse gases (CO_2) like other heterotrophic organisms. Thus, a completely clean process based on photosynthesis can be achieved using the two most important resources of our planet: water and sun. This photo-production of hydrogen is still at the experimental stage.

...from water, nuclear energy, biomass or photosynthesis

Hydrogen can be produced from water by electrolysis.
This hydrogen production method does not emit greenhouse gases or any other pollutant if the electricity used comes from a clean source, but it is relatively costly because it requires significant amounts of electricity.
It is also possible to dissociate water directly without electricity at high temperature through a thermochemical process, using the heat produced by a nuclear reactor.
Another method, still at the experimental stage, relies on unicellular green algae to produce hydrogen directly from solar energy.

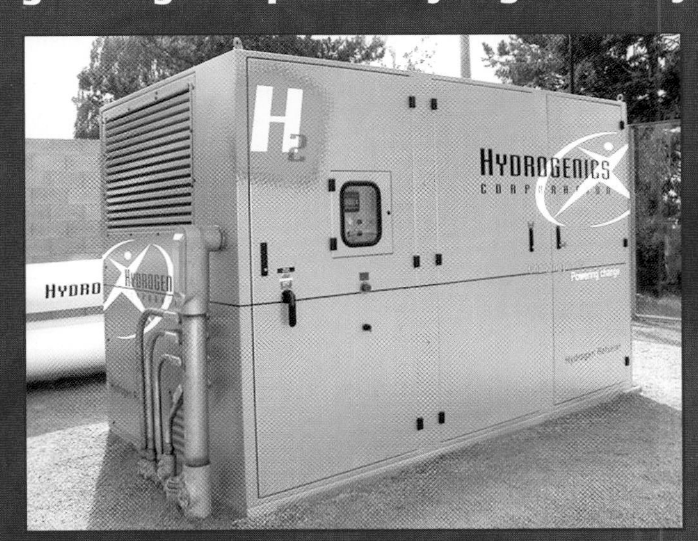

Proton exchange membrane (PEM) electrolysis is a new approach. A promising technology, it allows for a more compact design, it is simpler to operate, and it has fewer corrosion issues. However, PEM-based electrolysis is expensive at present because of the high cost of the membrane and the required use of noble metal electro-catalysts. Further improvements of the PEM coming from the fuel cell industry will benefit this technology.
High temperature electrolysis using water vapour at temperatures of 700-1,000 °C has also generated some interest. In this process, a significant part of the dissociation energy comes from thermal energy, obtained at a cost lower than electricity. The electrode is an O^{-} ion conducting ceramic, functioning in reverse operation as a SOFC. This process could be used in conjunction with HTGR (High Temperature Gas cooled Reactor) nuclear reactors or with solar concentrators.

Hydrogen...

Solar, wind, hydraulic and geothermal energies[1] do not directly cause pollution or emit greenhouse gases. They are readily available, essentially inexhaustible and can be converted into electricity, the most widely used energy vector. Thus:

- Solar radiation can be converted directly into electricity using photovoltaic cells.
- Mechanical energy from the wind can be used to produce electricity by wind turbines coupled to electric generators.
- Hydraulic turbines can, in conjunction with alternators, be used for large-scale production of electricity.
- Geothermal power plants can convert the heat from the depths of the earth into electricity with appropriate turbines.

The electricity produced from renewable sources, although easily distributed through grids, cannot be conveniently stored and must generally be used as soon as it is produced. Furthermore, the production of electricity from these sources is not constant, since renewable energy sources are inherently transient. Depending on the source, it is unavailable at night, when the wind subsides or the water level is at its lowest. Hydrogen can be used to store electricity as long as necessary by using the excess power generated during periods of low demand to produce hydrogen by electrolysis. When needed, it can be burned to produce heat, converted into electricity with a fuel cell for stationary applications, or be used to power vehicles. This can be done without emitting pollutants or greenhouse gases, as the only by-product is water[2]. Hydrogen thus respects the inherent qualities of renewable energy: pollution-free, and emitting no greenhouse gases.

Even if demonstration projects (such as the hydrogen village in Toronto, Iceland and the Island of Utsira in Norway) showcase the use of hydrogen in renewable energy systems, it is not yet used commercially. Despite issues pertaining to the overall efficiency of hydrogen-renewable energy systems, due to the chain of operations involved in storing and converting energy, the possibility of a clean energy vector with no impact on the environment or public health is a determining factor in its favor.

An especially interesting way to store renewable energy would be to concentrate solar radiation with mirrors to generate high temperature heat, which could then be used to produce hydrogen through the thermochemical dissociation of water. Although still at the experimental stage, this production method could be interesting in areas with long exposures to the sun. Another method of producing hydrogen from the sun is photolysis. This process is based on the use of photoelectrochemical cells, which, when subjected to solar radiation, can directly decompose water into oxygen and hydrogen. These cells are actually a combined system of photovoltaic and electrolytic devices.

Hydrogen can store renewable energies easily and can thus help to offset the main drawback of using renewable energies.

[1]Biomass should be added to these renewable energy sources. However, it cannot be exploited directly using a simple energy conversion device, and its renewable nature only persists on a timescale ranging from a year to a few decades at most.

[2]And possibly some NOx during direct combustion.

...a partner for renewable energies

Renewable energy sources, once converted into electricity, can be stored as hydrogen through water electrolysis. The hydrogen thus produced can then be stored and used when needed to power various applications, or for heating.

Compression, adsorption and absorption...

Compression storage of hydrogen

Compressed hydrogen is stored in cylindrical vessels made of low strength steel to prevent embrittlement. Because of the low density of hydrogen, these vessels only offer a mass to volume ratio of 14 kg/m^3 when filled at a pressure of 20 MPa.
Notable improvements have been achieved using composite materials. The design of this type of reservoir involves a sealed envelope, a resistant structure in wound fibre, covered by a protection layer. The basic structure is generally made of carbon fibres glued with resin. It is protected from collisions and environmental conditions by an external layer in glass fibres and resin. The shape of the reservoirs generally consists of a cylinder caped with two hemispheres.
The current standard service pressure rating is 35 MPa, but recent improvements are aiming at a service pressure of double this value (70 MPa), corresponding to an optimal value of the compressibility curve of hydrogen. The storage ratio (by weight: mass of H2/mass of Vessel) of cylinders ranges from 7.5 to 8.5% for commercial vessels operating at 35 MPa and 6.5% for reservoirs operating at 70MPa (which have the advantage of smaller volumes). Prototype pressure vessels have been designed that have achieved a storage density of 13%.
Automotive applications of these reservoirs allow the storage of 3 to 5 kg of hydrogen for a vehicle equipped with a fuel cell, corresponding to a range of 300-500 km. In the case of heavy vehicles, the mass of hydrogen required to achieve comparable autonomies can be 5 to 10 times greater. To ensure fast fill of less than 5 minutes for a light vehicle and 30 minutes for a heavy one, fuelling stations need to be able to dispense large quantities of hydrogen at high pressure.

Gaseous hydrogen adsorbed in porous materials

Adsorption of a gas by a solid surface is a physical process by which the density of a gas is enhanced close to the surface by intermolecular interactions between the solid and the gas molecules. Adsorption increases with gas pressure and decreases rapidly with increasing temperature. Because it is a non-chemical physical process, adsorption is reversible. It diminishes when pressure is lowered or when temperature is increased. To use this phenomenon to store gases requires that the largest possible number of gas molecules be exposed to a solid surface. Ideal materials are solids with large active surface areas, implying that the material needs to have an extensive porous structure. The material most commonly used, activated carbon, is made of microcrystals of graphite intermeshed in an extensive network of nanopores (pores with diameters of the order of nanometers). These pores have a considerable active surface that can reach many thousands of square meters per gram of carbon. Numerous studies have been performed to measure the adsorption properties of porous carbon materials (activated carbon, carbon nanotubes...). The mass of hydrogen that can be stored by adsorption processes in carbon materials such as nanotubes represents, according to molecular modeling, 1 to 2 % of the mass of the material depending on the pressure. Experiments have measured storage ratios ranging from 1 to 6%. Improvements may be achieved by developing new nanoporous materials better suited for hydrogen. At low temperatures, where the adsorption process is strongest, the storage is substantially improved by a factor of two at 77 K.

Hydrogen absorbed in metals

Absorption is the reversible chemical combination of hydrogen atoms with metals or metal alloys, resulting in the formation of metal hydrides. Certain pure materials such as vanadium, palladium or intermetallic compounds (yttrium, zirconium or a lanthanide with a transition metal) are known to reversibly absorb large quantities of hydrogen. Their storage capacity is often such that the quantity of hydrogen present in one cubic centimeter of a hydride can surpass that found in the same volume of liquid hydrogen. However, the mass of hydrogen absorbed expressed as a percentage of the mass of the absorbent, lies between 1.5 and 7.6%. The volumetric storage densities are of the order of 0.13 g/cm^3.
The fact that these metals and alloys are heavy and costly hinders their use on a large scale. Moreover, the hydrogen absorbed must be very pure in order to maintain absorption capacity without degradation of the material after repeated cycles of absorption and desorption. Finally, from a practical point of view, the heating effects associated with hydruration (absorption or filling) and dehydruration (desorption or emptying) must be taken into account. Absorption is an exothermal process (~150 kJ/kg) so the heat produced during filling must be evacuated. Desorption is endothermic and requires the addition of heat. The temperatures of the hydruration reactions typically range from 300 to 650 K, at pressures of the order of 0.1 MPa. Onboard vehicles, the heat produced by the engine can be used to desorb hydrogen from the hydrides. Some degree of cooling of the tank may be required during filling.

...Storing hydrogen as a gas

Compressed hydrogen gas at 20 MPa is stored in steel cylinders of 50 litres. Highly resistant composite material cylinders can be used to store hydrogen at a pressure of 35 MPa. It is possible today to increase the pressure to 70 MPa, almost doubling the quantity stored.
Gaseous hydrogen can also be stored by adsorption: the hydrogen molecules are then attracted by the microporous surface of activated carbon or carbon nanotubes. The efficiency of the procedure remains to be shown.
Hydrogen gas can also be stored by absorption in a metal or a metal alloy. During this process, hydrogen diffuses into its structure and combines in a reversible manner with the metal atoms, forming metal hydride. While making it possible to store large quantities of hydrogen, this storage technology has decreasing efficiency with use. In addition, metal hydrides are costly and heavy. Furthermore, new hydrogen storage technology currently being investigated involves chemically combining hydrogen with nitrogen or with boron and sodium.

Chemical hydride

Sodium borohydride ($NaBH_4$) can, in the presence of a cobalt-based or ruthenium catalyst, react with water, producing water and sodium borate. One has only to put the borohydride-water solution in contact with the catalyst to obtain hydrogen.
Borohydride is not toxic or flammable. It is easily handled. This way of storing hydrogen could be particularly interesting for automotive applications, especially because the hydrogen produced is pure and does not contain carbon monoxide, which can poison the catalysts used in PEM fuel cells. The cost of the catalyst, however, is high and issues pertaining to the spontaneous emission of hydrogen and the recycling of sodium borate must be addressed.

Liquefaction...

At atmospheric pressure, hydrogen becomes liquid at 20.3 K. As a liquid, hydrogen is much denser (70.8 kg/m^3) than as a gas under normal conditions (0.09 kg/m^3). The latter would have to be compressed to 182 MPa to obtain the same density. Hydrogen at 70 MPa has a density 1.8 times lower than liquid hydrogen. There is thus a great deal of interest in storing and transporting hydrogen in liquid form. However, a certain level of cryogenic engineering is required to liquefy it and maintain it in this state.

Liquefaction

Hydrogen can be liquefied using the Claude cycle. Gaseous hydrogen is first pre-cooled in a liquid nitrogen heat exchanger. It is then subjected to a succession of compression followed by expansion cycles, producing work during which the energy of the hydrogen gas is progressively extracted. The last stage is to subject the gas to a Joule-Thompson expansion without exchange of work or heat with the environment, which produces a cooling effect as long as the temperature of the gas is below a critical value.

Other liquefaction methods rely on heat exchanges of hydrogen and pre-liquefied helium (helium and neon mixture) or on the magnetocaloric effect. The latter, still at the experimental stage, relies on the fact that magnetic materials release heat when subjected to a magnetic field, and lose heat when suppressed.

The energy necessary to attain the liquid state, depending on the quantity of hydrogen, is two to ten times greater than the energy required to compress it to 70 MPa, specifically 22 MJ/kg.

Cryostats

As any cryogenic liquid, liquid hydrogen is kept in "cryostats", containers with two walls separated by a vacuum to impede direct conduction of heat through the container. In addition, a polished silver coating protects from radiative heat transfer by reflecting radiation. Generally made of glass in the past, cryostats are now mostly made of stainless steel, and have a storage capacity ranging from several litres to tens, hundreds or even thousands of cubic meters.

The thermal insulation of these containers, as good as it can get, can never be perfect. Thus there is a low but constant level of boil-off of the liquefied hydrogen. Cryostats are designed to allow the release of hydrogen vapour to prevent dangerous build-ups. Boil-off represents a loss of 1% of the total mass of the liquid per day. In the case of automotive applications, the vehicle should not remain in a confined area for more than two or three days even if the reservoirs are capable of resisting pressures of 0.5 to 0.8 MPa. For stationary applications, the released vapour can be captured by cryoadsorption on activated carbons in pressure resistant enclosures, which, when brought to ambient temperature, become compressed gas containers.

On-board uses of hydrogen

Liquid hydrogen has been used in certain hydrogen-fuelled prototype vehicles. Although liquid hydrogen reservoirs require less space than compressed hydrogen, the combined weight of the storage system and the fuel is not better than compressed storage (8%). There might be an advantage in using liquid hydrogen for heavier vehicles, where the volume issue might not be as critical. In addition, because of the evaporation of liquid hydrogen, a liquid hydrogen vehicle could not be left in an enclosed area for two or three days without taking specific precautions, even with storage tanks rated to withstand pressures of 0.5 and 0.8 MPa above atmospheric conditions.

Liquid hydrogen has been widely used in aeronautics to minimize the storage volume requirements. The mass of hydrogen required to provide an equivalent amount of energy represents 35% of the mass of kerosene; volume requirements are four times larger.

Liquid hydrogen is essential for satellite launchers. It has been in use for over 50 years as fuel for rockets. The volume and the mass of the various stages of the rocket must be minimized. The ratio of the mass of the payload to the reservoir is 300% for a small stage and 600% for the main stage of Ariane 5. This is due to the size of the shell of the reservoirs, which contains tens to hundreds of cubic meters of liquid hydrogen, and to the fact that the hydrogen is only needed for the flight duration (about half an hour). This allows the use of lightweight insulation for the tank since it can be filled at the last moment.

...to store more hydrogen

Hydrogen in its liquid form is 800 times denser than in its gaseous form under ambient conditions. Liquid hydrogen would be very easy to store if its temperature did not have to be brought down to -253 °C, a process that requires a large amount of energy and special cryogenic equipment to store and handle the liquid. This storage technique is complex and costly and is only justified for storing the large quantities required in aerospace applications such as rockets. It could also be used in large vehicles, trains and boats or to store large quantities at production facilities.

Transporting hydrogen...

Compressed hydrogen cylinders

The transportation of compressed hydrogen cylinders is primarily carried out by trucking. Because of the low efficiency of hydrogen storage as compressed gas in cylinders (due to the low density of the hydrogen and the weight of the high-pressure vessels), a 40-ton truck can transport a mass of hydrogen six times smaller than methane and nearly eighty times less than gasoline. Moreover, the energy necessary to transport this hydrogen over a distance of 500 km would be equivalent to the energy content of the hydrogen being moved – not a very efficient process! Transportation by railway or ship would decrease this slightly but at the price of more complex filling and unloading procedures and of a less flexible distribution infrastructure.

Gas pipeline networks

The intensive industrial use of hydrogen in the process industry really began in 1938 in the Ruhr valley when a 240 km long pipeline was built, capable of transporting 250 million normal cubic metres of hydrogen per year. Currently, Western Europe and the United States have hydrogen pipeline networks of 1500 km and 900 km long respectively. Other smaller installations also exist in South America and in Thailand. These pipelines are made of traditional steel pipes 100 to 200 mm in diameter. They carry hydrogen compressed to a pressure of 0.34 to 10 MPa, linking production units (typically producing hydrogen through steam methane reforming, ethylene cracking or by chloro-alkaline electrolyzing) to industrial facilities. Pipeline distribution networks of Air Liquide can be found in France, Northern Europe, and the United States. Other networks of Linde, Air Products and Praxair also exist in Britain, America, Brazil and Thailand.

Hydrogen used as an energy source could be distributed on a large scale by using and extending the networks mentioned above. It would also be possible to adapt natural gas pipelines to distribute a mixture of natural gas and hydrogen in a proportion of 9/1. Distributing larger quantities of hydrogen would require re-sizing the pipes and modifying the compression equipment.

Liquid hydrogen

Liquid hydrogen is transported by trucks in cryogenic reservoirs or better yet by cryogenic tanker trucks, which can carry liquid hydrogen more efficiently than compressed gas. Indeed, five times more hydrogen can be stored in liquid form than as compressed gas under the same conditions. Transport of liquid hydrogen by rail or by ship would be even more efficient.

...in bottles or by pipelines

Hydrogen is transported by truck, rail or water using either compressed gas cylinders or cryogenic vessels. Due to the low density of hydrogen, this operation is costly and inefficient. There are better options: gas pipelines can be used to distribute gas at pressures of 10 MPa.
This method is currently being used to transport hydrogen for the chemical industry from production areas to consumer sites.

Distributing hydrogen...

Distributing hydrogen through service stations would make it available in a safe, convenient way, as service stations currently do with automotive fuels. But the existing distribution system cannot be easily adapted to hydrogen because gasoline is liquid, while hydrogen is most commonly used as a gas. The fuels distributed today are obtained by refining petroleum from buried oil reserves. Hydrogen, on the other hand, can be produced almost anywhere from all sources of energy - fossil, nuclear or renewable. Hydrogen can be made at the distribution centre or at a central production site from which it can be transported to distribution sites by road, rail, water or pipeline. The installations required for distributing hydrogen depend on the type of storage used onboard the vehicle – compressed gas, liquid etc.

Vehicles using compressed hydrogen can be refueled in the following ways:

- The direct transfer of pressurized gas from a large storage system, currently performed at 35 MPa.
- Reservoir exchange. An empty reservoir would be replaced with a pre-filled reservoir at the station. This solution would require a system that makes the exchange easy and faster than refilling the tank onboard.

There are two methods for transferring hydrogen from one cryostat to another. First, the container to be emptied is tightly sealed and connected to a source of pressurized hydrogen gas. A tube is then inserted into the liquid and connected to the receiving vessel. The second method uses a cryogenic piston pump.

Likewise for liquid hydrogen vehicles, the system would involve either:

- Direct transfer from a large reservoir.
- Tank exchange. This system, though not tested, has, in addition to allowing for quick refueling, the advantage of keeping the handling of the liquid hydrogen out of the hands of the customer.

Efficiency of hydrogen distribution systems

- **Production:** Producing hydrogen from steam reforming of fossil fuels (natural gas, petroleum products) has an overall efficiency of up to 90%. The best efficiency of hydrogen production from water by electrolysis is between 75% and 85%.
- **Conditioning:** Compressing hydrogen gas leads to a loss of 10 to 15% of the energy content if hydrogen is stored at a pressure of 20 to 80 MPa. In liquid form, the energy required for liquefaction corresponds to a loss ranging from 150% for small scale liquefiers (several kg per hour) to only 30% for large scale liquefiers (more than 1 ton per hour).
- **Transportation by road:** Due to the weight of the reservoirs, the energy required to transport hydrogen over a distance of 500 km is equal to the energy content of the hydrogen being delivered. It is more efficient to transport compressed hydrogen through gas pipelines since the energy cost for transporting over a distance of 150 km is only 1.4% of the energy content of the hydrogen being distributed. The cost for this energy stems from the high-pressure pumps required to ensure a constant supply of gas as it is transported by the network.
- **On-site production of hydrogen at refueling stations:** The efficiency of this process is a function of the number of vehicles serviced per day. The energy loss varies from 75% for 100 vehicles per day to 40% for 2,000 vehicles per day, the quantity of hydrogen needed being 1,700 kg and 34,000 kg per day respectively. An electric power supply of 5 to 81 MW would be required.

Efficiency of hydrogen distribution systems

Hydrogen must be transported over the smallest distance possible. For large quantities of hydrogen, transportation in liquid form could be done by rail or by water. However, pipeline distribution represents the better solution. Production on site by electrolysis in small units requires a lot of energy, preferably from renewable energy sources.

...at service stations

A hydrogen refueling infrastructure is essential to hydrogen vehicles. Hydrogen refueling stations will be able to:

- **distribute compressed hydrogen by a high pressure refueling system;**
- **exchange empty containers of compressed gas for full ones (similar to the exchange of butane or propane tanks);**
- **distribute liquid hydrogen by a transfer line or by exchanging cryogenic reservoirs.**

These stations would be supplied with compressed or liquid hydrogen by trucks or through pipelines. They could also directly produce hydrogen by electrolysis of water using electricity from the grid or produced from wind turbines or photovoltaic cells, or by reforming biomass.

The safe use of hydrogen...

Hydrogen can burn or even explode in oxygen or air when ignited. Like any other fuel, hydrogen must be handled with care. The following tables summarize its combustion properties compared to propane.
An explosion can occur if enough hydrogen is present to form a flammable cloud. An explosion is a sudden and brutal release of energy, which creates a flame front and an overpressure wave. It can occur in one of two forms:
• Deflagration occurs typically when the concentration of hydrogen in air is between 4% and 18% or 59% to 75%. In a deflagration, the flame front moves at subsonic speed, the fresh gas being compressed by the expansion of the burning front which can result in a progressive increase of the pressure wave.
• Detonation typically occurs when the concentration of hydrogen is between 18% and 59%. In a detonation, the flame front moves at supersonic speed with the overpressure wave, forming a shockwave.
Because of the flammable and explosive nature of hydrogen, leaks represent a serious risk. Hydrogen, being the lightest and smallest of molecules, has a strong tendency to leak regardless of the storage system used to contain it (compressed gas, pipelines, liquid hydrogen container, etc). Small leaks in unconfined areas disperse quickly, preventing the formation of a substantial flammable cloud. On the other hand, large leaks in confined areas represent a severe risk. These risks can be reduced through proper ventilation and management of possible areas where ignition could occur.

The new 70 MPa hydrogen reservoir previously mentioned, if designed properly, offers a safety level comparable to classical storage tanks of gasoline or liquefied petroleum gas (lpg). Generally, the risk of leaks must be managed during the design of a fixed or mobile hydrogen system, and proper safety procedures, involving leak detectors and ventilation, must be planned.

	Units	Hydrogen	Propane
Flammability range in air	% vol	4 – 75	2.1 – 9.5
Minimum ignition energy (at stochiometry)	MJ	0.02	0.26
Auto-ignition temperature in air	K	858	760

Other properties of hydrogen from the perspective of safety	
Adiabatic flame temperature of hydrogen at 300 K	2 318 K
Detonation limits (vol %)	13-65
Theoretical explosion energy (in kg of TNT/m3 of gas)	2.02
Detonation overpressure at stochiometry	1.47 MPa
Diffusion constant in air	0.61 cm^2/s
Flame velocity in air	260 cm/s
Detonation velocity in air	2.0 km/s
Stochiometric composition in air (vol.)	29.53%

...The risks of hydrogen

Hydrogen is easily ignited in air. Since it is very light, it usually disperses very quickly and only presents a real danger when it is maintained in a confined space without proper ventilation. It could even explode when a cloud of hydrogen and air is formed, but this is unlikely with proper ventilation.
The safe use of hydrogen requires avoiding leaks, which implies careful monitoring with detectors and the use of a proper and effective ventilation system. Proper design features, such as mounting hydrogen tanks on roofs should also be considered.

Flame from a hydrogen reservoir

Flame from a gasoline tank

The Hindenburg accident: a public relations nightmare for hydrogen

In 1937, the German airship Hindenburg[1], arrives in New Jersey during a thunderstorm after having crossed the Atlantic. It catches fire and is destroyed, killing 35 of its 100 passengers. The hydrogen contained in the airship was blamed for the accident. Investigators concluded that the tragedy occurred because of the flammable paint covering the airship, and that despite the fact that hydrogen was ignited during the incident, hydrogen was not responsible for the deaths of the victims.

[1]Constructed by the German firm Zeppelin.

The hydrogen economy...

In order for hydrogen to play an important role in the transportation sector and elsewhere – at home, in industry or in agriculture – the development of a large-scale infrastructure is required. This involves the implementation of an efficient production and distribution infrastructure, and of all the technology needed for its safe use as an energy vector. It also implies considerable investments and the creation of a vast market: the birth of a hydrogen economic system.

The production of large quantities of hydrogen could be realized using most of the existing infrastructure for fossil fuels. This would, however, require capturing and sequestering the carbon dioxide emitted ideally at production sites in order to minimize the risks to the environment.

A centralized or distributed production infrastructure based on biomass, nuclear energy or other renewable energies would essentially have to be created from the ground up. A distribution system would also have to be established, and energy conversion devices for hydrogen would have to become available. Important decisions will have to be made concerning our future energy system, from the economic and financial perspectives.

While an energy system based on hydrogen represents a straightforward evolution for some and pure science fiction for others, some believe it could create a new global society without carbon, a hydrogen civilization. Iceland is currently trying to make this vision a reality. Through a project launched in 1999, Iceland is striving to become the first society to build a hydrogen economy. A first step will consist in converting their fleet of automobiles, buses, heavy trucks and trawlers to hydrogen. The second step will be to use hydrogen for lighting, heating as well as industrial and domestic applications. Geothermal energy would be used to produce hydrogen.

A similar project, although at a much less advanced stage, has been proposed for Hawaii, using geothermal and solar energy.

...pathways to commercialization

A hydrogen economy would involve extensive use of this energy vector for transportation, heating and lighting, as well as for industrial and domestic activities. It would not only involve installing production units and implementing a distribution network, but also the production and commercialization of all the equipment necessary for its use: basically creating a hydrogen market.

Hydrogen

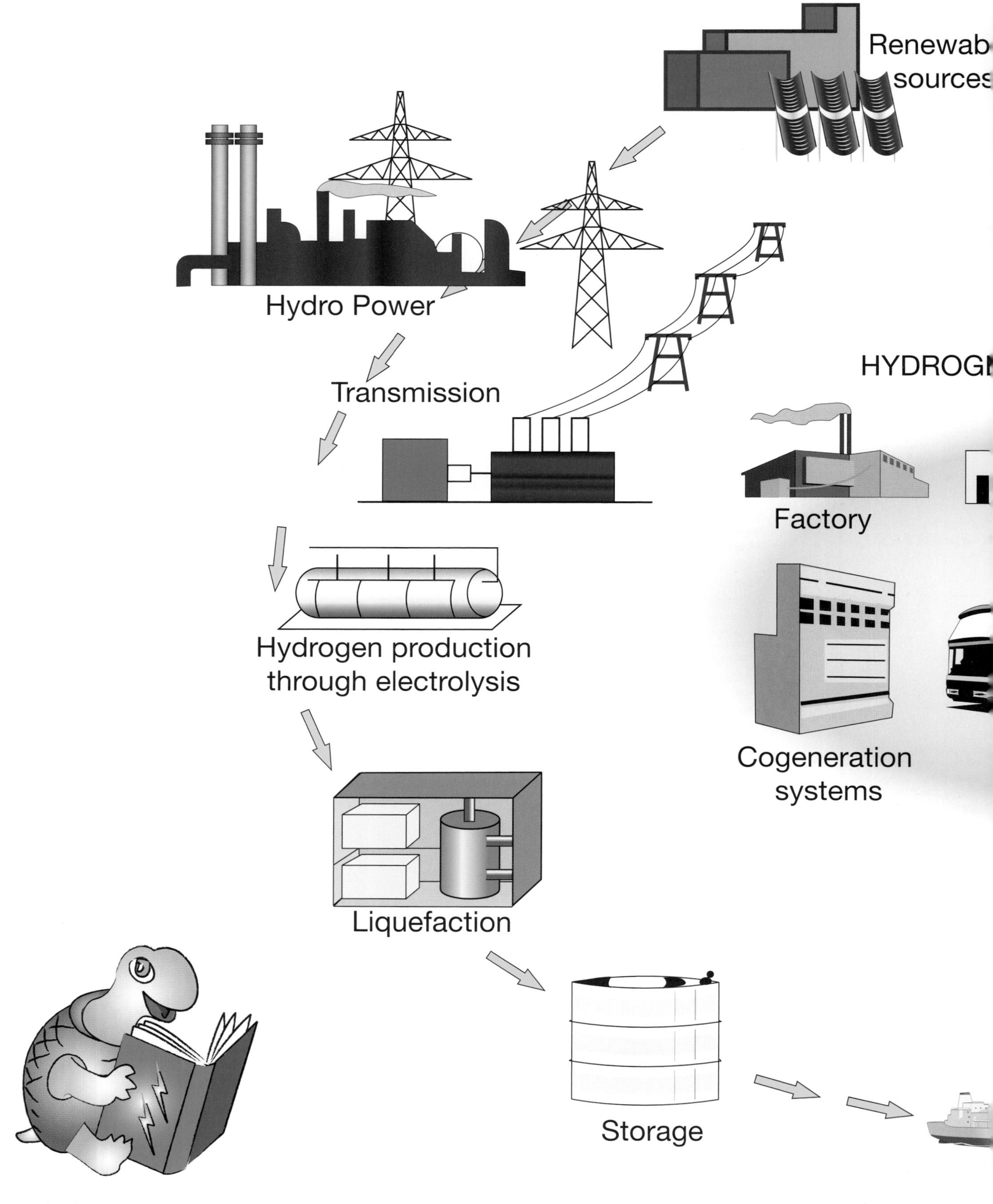

Hydro Power
Transmission
Hydrogen production through electrolysis
Liquefaction
Storage
Factory
Cogeneration systems

economy

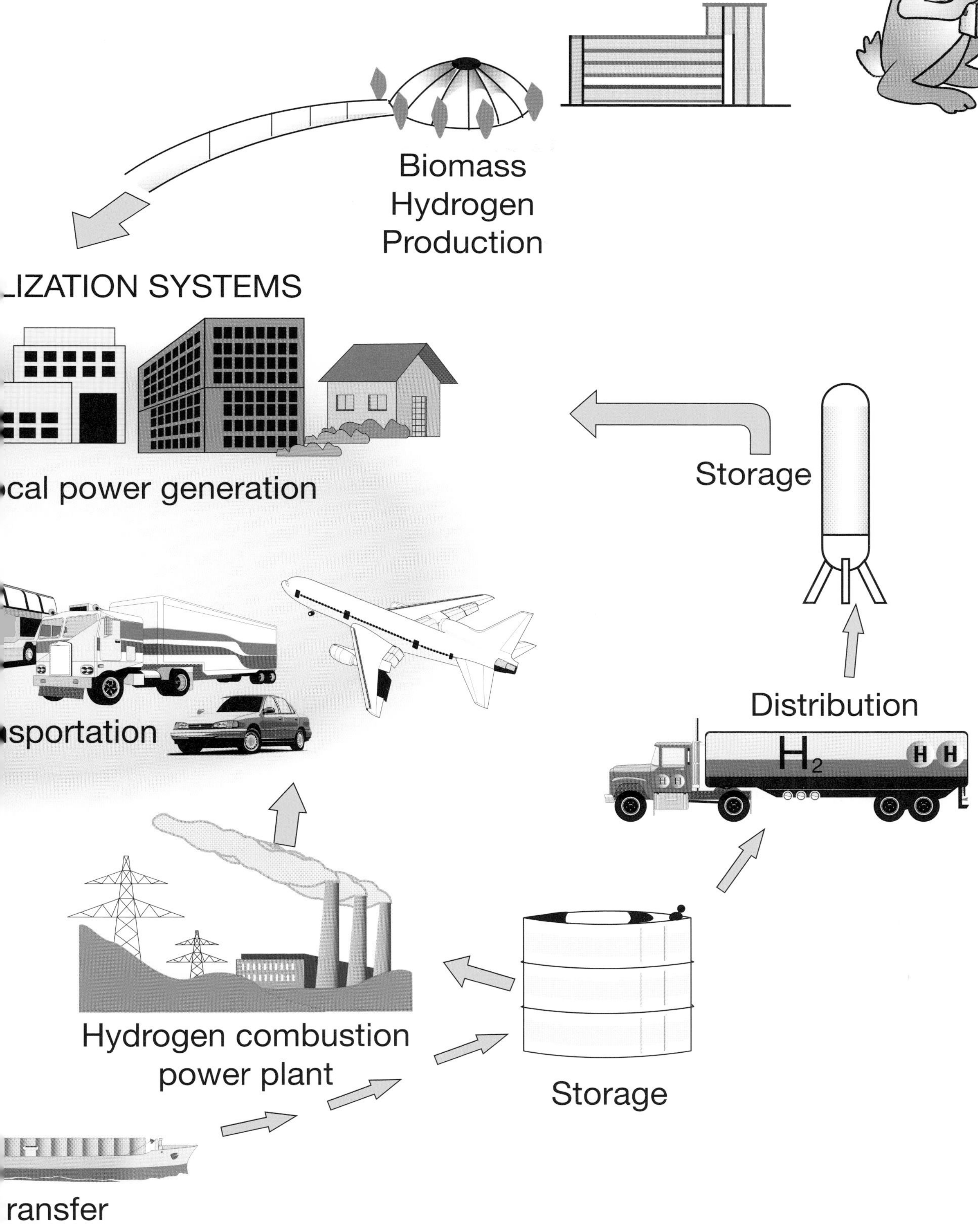

Biomass Hydrogen Production

LIZATION SYSTEMS

cal power generation

Storage

sportation

Distribution

H_2

H H

Hydrogen combustion power plant

Storage

ransfer

What can be done...

Hydrogen could help preserve the environment and provide an answer to the depletion of fossil energy sources. However there remain several obstacles, technical and economic, to the introduction of hydrogen into the energy market.

Producing hydrogen from fossil sources would generate significant quantities of CO_2, which would have to be safely sequestered. Large scale production using nuclear energy through the dissociation of water either by high temperature electrolysis or thermochemical processes could be considered. Renewable energy from various sources could be used for the decentralized production of hydrogen for various applications, particularly off-grid. Progress is still required to achieve acceptable costs for such systems, but **decentralized production and use of hydrogen may be one of the most significant contributions of hydrogen to our future energy system**.

The storage issue could be solved either by using (i) liquid hydrogen for large scale use in stationary applications or for large vehicles, boats or airplanes or (ii) gaseous hydrogen compressed to high pressures for use on a smaller scale, for example in light vehicles. More research and development activities are required to store hydrogen in metal hydrides or on carbon nanostructures on a large scale.

The transportation of gaseous hydrogen by pipelines or of liquid hydrogen by boat or train would be most cost effective. Hydrogen should be transported over the shortest distances possible. This could be achieved through distributed production which would require efficient, low volume production facilities.

Fuel cells, used to convert hydrogen into electricity have been considerably improved over the last few years. Their deployment, however is hampered by some remaining technical issues, but particularly by their cost, which is still too high.

The cost of hydrogen delivered at a service station remains four to six times higher than oil. Lower costs for hydrogen can be expected from technological advances, while higher costs for oil are expected as resources are depleted and the costs of extracting oil become higher. Hydrogen use becomes more interesting if the externalities are taken into account. Externalities are the costs associated with public health and the environmental impact of fossil fuels.

Finally, the large scale use of hydrogen and its production from various sources have drawbacks that must be mitigated or avoided. For instance:
- Hydrogen released into the atmosphere through the unavoidable small leaks may have unforeseen impact on the environment;
- The long term geological sequestration of CO_2 arising from the production of hydrogen using fossil fuels may impact the environment either through leaks to the surface or by modifying the nature of geological layers because of its acidity;
- Using renewable energy to produce hydrogen may have an impact on the quality of life of local communities (such as noise from wind turbines, the aesthetics of solar panels);
- There are issues pertaining to the use of biomass to produce hydrogen: delays in sequestering the CO_2 by absorption through photosynthesis, the fact that certain plants only absorb CO_2 at the beginning of their growth and release it at the end, biomass requires large surfaces and important quantities of fresh water, the growth of biomass has an impact on the environment through pollution arising from the use of fertilizers and pesticides, and the impoverishment of soils.

...to use hydrogen as a fuel

The large scale use of hydrogen would require:

- **Producing hydrogen without releasing carbon dioxide, either from renewable energy sources or nuclear power or by sequestering the carbon dioxide released during the reforming of fossil fuels;**
- **Decreasing production costs of hydrogen;**
- **Improving the storage density of hydrogen and its safe handling;**
- **Developing distribution networks adapted to different production modes and to the needs of end-users;**
- **Decreasing the cost of fuel cells and making them more reliable;**
- **Exploring and addressing the possibility of certain environmental issues that could arise from unavoidable hydrogen leaks, as well as addressing the safety and the environmental impact of sequestering carbon dioxide over the long term;**
- **Considering the impact of using renewable energy on a large scale on the quality of life of local communities (such as noise from wind turbines, the aesthetics of solar panels).**

What is being done...

In Canada:

Different initiatives are currently under way, such as the hydrogen highway (8 to 10 service stations to be put in service for the 2010 winter Olympic games in Vancouver), the Hydrogen village in Toronto and the HyPort demonstration project at the Montreal Airport. A Canadian roadmap for the hydrogen economy has been proposed by Natural Resources of Canada in collaboration with the Canadian Hydrogen Association.

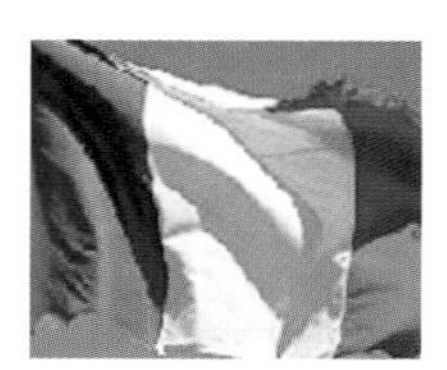

In France:

Research and development activities in the field of energy technologies have benefited from an increased interest in recent times. Research activities in hydrogen storage and fuel cells have developed considerably. A Hydrogen National Action Plan (PAN'H) was launched in 2005. It is an innovative research project combining the public and the private sectors to develop an industrial network of hydrogen directed primarily towards the automobile sector.

In European Union:

A technology platform for hydrogen and fuel cells was launched in January 2004. Current urban transportation projects include the Clean Urban Transport for Europe (CUTE) and the Ecological City TranspOrt System (ECTOS). The"Quickstart' programme includes the "Hypogen" project for the production of hydrogen from fossil resources with CO_2 sequestration and the "Hycom" project on vehicles using hydrogen produced from renewable energy.

In Iceland:

The Icelandic program previously discussed proposes to convert the whole of the economy of the island to hydrogen by eliminating fossil fuels within twenty years using geothermal energy, an important natural resource of this volcanic island.

In the United States:

The development of hydrogen fuel cell vehicles is part of the "Freedom Car & Hydrogen" initiative. The year 2015 is considered a first step towards the establishment of a hydrogen economy, when a decision on the commercialization of hydrogen technologies is expected to be made. From 2015 onwards, an infrastructure investment phase will begin to put into place a "hydrogen economy" by 2025. The objectives for the number of fuel cell vehicles are 250 by 2005-2008, 2500 by 2009-2011, 25,000 by 2012-2014 and 50,000 by 2015-2017.

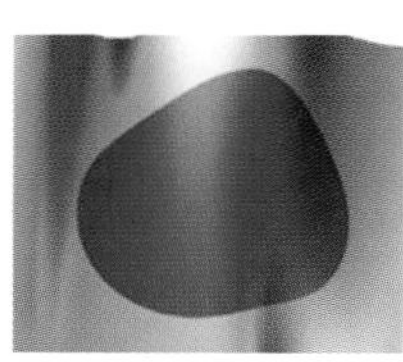

In Japan:

The objectives of the Japanese government are to have 2.2 GW hydrogen generators installed for stationary applications and 50,000 vehicles on the market for 2010. The year 2005 was seen as a key stage towards the introduction of the technology on the market. The industrial production of hydrogen technologies is expected to begin by 2010. The Japanese hydrogen project World Energy NETwork (WE-NET) is part of a broader program, the New Sunshine project, managed by the New Energy and Industrial Technology Development Organization (NEDO). It consists of three stages:

- The first stage, completed, involved research activities on the production, storage and use of hydrogen.
- The second stage, in progress, should lead to construction of equipment for the production, storage and use of hydrogen.
- The third stage, the longest, aims to deploy the hydrogen technologies developed. It is projected to last from 2006 to 2020, at which point the technologies are expected to be widely used.

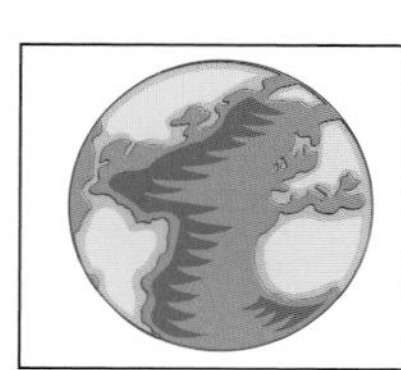

Worldwide:

The International Partnership for Hydrogen Economy (IPHE) was created in 2003. Grouping most of the industrialized countries, including the European Union, its purpose is to accelerate the introduction of hydrogen on the energy market.

...Showcasing and developing Hydrogen

In addition to Iceland, which is attempting to establish a hydrogen economy, every part of the world has its own hydrogen program.

In France: Action plan national (PAN' H) launched in 2005.

- In Europe: A technology platform for hydrogen and fuel cells was launched recently, in addition to various urban transportation projects.
- In Canada: A Canadian roadmap to a hydrogen economy has been proposed and several demonstration projects are under way, such as the Hydrogen Highway, the Hydrogen village in Toronto and the HyPort demonstration project at the Montreal Airport.
- In the US: The development of hydrogen fuel cell vehicles is part of the "Freedom Car & Hydrogen" initiative; other programs have also been set up.
- In Japan: The "World Energy NETwork" project is being developed until 2020.
- Worldwide: The International Partnership for a Hydrogen Economy (IPHE) has been launched.

Hydrogen and electricity are complementary energy vectors that can transform all sorts of energy sources without producing

Conclusion

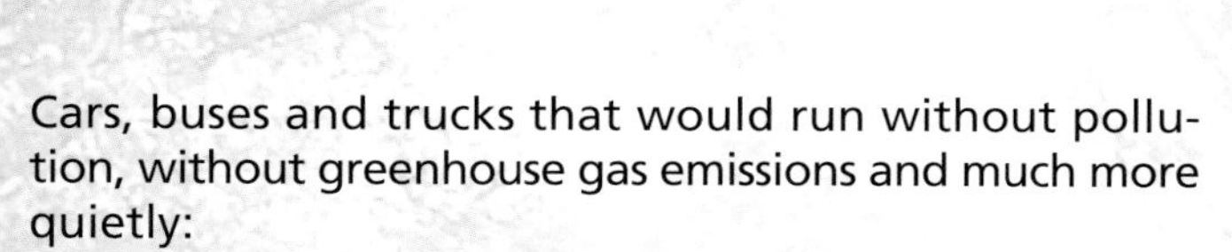

Cars, buses and trucks that would run without pollution, without greenhouse gas emissions and much more quietly:

- *It could be possible with hydrogen thermal engines or electrical engines powered by hydrogen fuel cells.*

Cell phones and laptops that would work longer and that could be recharged by inserting a simple cartridge:

- *It could be possible with micro fuel cells fuelled by hydrogen or methanol.*

Using the power from the sun, the wind or from rivers for heating and lighting, even at night, when rivers have dried out or the wind has subsided:

- *It could be possible with photovoltaic cells, wind or hydraulic turbines generating electricity that could be used by an electrolyzer to produce hydrogen, which would then be stored and used as a fuel later. Nuclear energy could also be used in the same way.*

However:

- **Several technological and scientific breakthroughs are required to ensure reliability and cost reduction of the equipment necessary for an efficient hydrogen-based energy system: production units, fuel cells, storage systems etc.,**
- **The reluctance of some segments of the general public and of the proponents of traditional energies must be overcome,**
- **A political will to preserve our environment and improve public health is required.**

Annex 1

What is energy?

Work is energy...

Work

The work required for a crane to raise an object is proportional to the weight of the object and to the height to which it is raised. Similarly, work must be provided to plough the ground, or to saw wood or metal. Work must be provided to a car in order to overcome the friction forces from the ground and the air, and maintain its speed. Work is provided by the pistons of the engine, whose motion is caused by the pressure generated by the combustion by-products of the fuel in the cylinders.

Kinetic energy

In order for a vehicle to acquire speed it must be pushed or pulled. The work performed on the vehicle to set it in motion is transformed into kinetic energy. Kinetic energy is the energy that any moving object possesses because of its motion.

Force

A force is essentially the source of any change in shape or velocity of a body. The fundamental principle of dynamics states that any body subjected to a force $\vec{F}$ undergoes an acceleration $\vec{\Gamma}$ which is proportional to this force and inversely proportional to its mass: $\vec{F} = M.\vec{\Gamma}$. A force of one Newton (symbol : N) is a force such that a mass of 1 kg acquires an acceleration of 1 m/sec^2: in other words its velocity changes by 1 m/s every second. The mass of an object should not be confused with its weight, which is the force exerted on it by the Earth's gravity. A mass of 1kg actually weighs 9.81 N.

Work

Work is defined as the product of the displacement of an object by the force exerted on it along the direction of this displacement. If the force is oblique, only the component along this direction of the object is taken into account. The unit of work is the Joule (symbol: J). One Joule is the work resulting from the action of a force of 1 N applied on an object as it is moved by one meter. This unit is small: a kiloJoule (1000 Joules) is more frequently used. All mechanical energies – kinetic, gravitational or elastic- are also measured in Joules, because they are either the source of or equivalent of Work.

Kinetic Energy

Kinetic energy (symbol : K) is the energy of an object in motion of mass M moving at a velocity v : $K = (1/2)Mv^2$. A mass of 1 kg moving at a speed of 5 km/hour would have a kinetic energy of nearly 1 joule. The kinetic energy of a one ton car moving at 100 km/hour corresponds to the same energy gained by a car that dropped from a height of 39 meters. The work caused by the friction exerted by air on the same car during 1 km is equal to 300 kJ.

Gravitation

The law of universal gravitation, discovered in 1684 by Isaac Newton, states that the force of attraction between any two bodies is proportional to their masses M_1 and M_2, and inversely proportional to the square of the distance between them: $F= G\ M_1.M_2/d^2$ where G is a universal constant, equal to $6.67.10^{-11}$ N.m^2/kg^2. This formula applied to a mass of 1 kg lying on the surface of the Earth (whose mass is $5.976.10^{24}$ kg and whose equatorial radius is 6,378 km) yields the value 9.80 N as the weight of the mass. In fact, the commonly used value is 9.8 N, which accounts for the small variation of the radius of the Earth and the centrifugal force associated with its rotation.

...mechanical energy

Gravitational potential energy

Any two bodies with a mass attract each other. This universal force of attraction, which is very small for light objects, becomes important for massive bodies such as the sun, the earth and the moon. On earth, gravity is nothing more than the action of our planet on objects in its vicinity, and weight is the corresponding force of attraction. When an object increases its velocity as it falls freely, the work exerted by its weight is transformed into kinetic energy. Any object raised above ground initially possesses a potential energy associated with the fact that it can fall. Thus, a load being hauled up acquires gravitational potential energy because of the work exerted by a crane. The water trapped by a dam constitutes a reserve of potential energy that can be used to power turbines coupled to electric generators.

Isaac Newton

Elastic energy

Elastic energy is a form of potential energy associated with the deformations of certain objects under stress. When the stress is no longer applied, this energy can be transformed into kinetic energy. For instance, a bow transforms its elastic energy into the kinetic energy of an arrow when the string is released.

Physical exertions such as practicing sports or a long walk eventually tire us. This is the way our body shows that our muscles have pushed, pulled, hauled or moved objects. Our muscles have exerted a force and caused a displacement, yielding work or mechanical energy.

Heat is energy

Heat, which we perceive as temperature, is a manifestation of the state of motion of the particles making up matter: vibration of atoms or molecules of solids or motion of the molecules of liquids and gases. The higher the temperature, the more intense these motions become, and the larger the kinetic energy of the molecules and atoms. At our scale, we perceive these effects as a rise in temperature when a body receives heat, and as a lowering when it loses heat. Heat also manifests itself through the changes in phases of a body: liquefaction or solidification, evaporation or condensation, depending on whether it receives or loses heat.

Heat is a form of energy

Heat can be transferred by contact through thermal conduction without changing its nature (through the transmission of the momentum of molecules, such as when water is brought to the boiling point on a stove). Heat can also come from the transformation of mechanical energy. This occurs, for instance, when brakes are applied to stop a car: the friction forces exerted by the break pads on the discs transform the kinetic energy of the vehicle into heat, which is transmitted to the environment. The kinetic energy lost by the car is transferred to the breaks as heat. This correspondence between heat and energy allows us to measure heat in Joules, the same units applied to mechanical energy.
Conversely, the heat produced by the combustion of coal, oil or gas can be converted into mechanical energy through a gas turbine, an internal combustion engine, or a jet engine. Thermodynamics, however, limits the efficiency of this conversion, which is only partial. The possibility of converting heat into mechanical energy is extremely important to our technology and is commonly used in many systems or appliances.

Temperature scales

A temperature scale allows us to determine the relative warmth of an object. Temperature is measured by a thermometer. It is not an extensive (or scalable) property of an object, but rather an intensive property, measured using a specific physical phenomenon to establish a reference scale. And as such, it is not a property that can be measured directly. The most common temperature scale is the Celsius scale, whose basic unit is the Celsius (C). The Celsius scale is defined using two reference points, the fusion and the normal boiling point of water, associated respectively with 0 and 100 C. The scale is uniformly divided into 100 equal segments. The most commonly used thermometers (such as those based on the dilatation of mercury or alcohol, or thermocouples and thermisters) are calibrated using the Celsius scale.
The absolute temperature scale is based on the fact that the pressure of a gas is a direct function of temperature[1]. Temperature cannot be lowered below a certain value, the absolute zero, the point at which all molecular motion should, in principle, come to a stop. This point is used as the origin of the absolute temperature scale, whose unit is the Kelvin (symbol : K) , which is defined in such a way that the triple point of water, a perfectly stable and unambiguously defined property of water, corresponds to a temperature of 273.16 K. This temperature scale coincides with thermodynamic temperature defined below (see the section on Carnot's principle). The Celsius scale is related to the absolute scale in the following way :

Temperature in C = Temperature in K – 273.15

Thus, a temperature difference of 1 K corresponds to a temperature difference of 1 C.

From the calorie to the Joule

When the correspondence between heat and mechanical energy had not yet been established, heat was measured according to its own specific unit. The calorie (symbol: cal) was then defined as the amount of heat required to raise the temperature of a gram of water at ambient pressure. A calorie corresponds to 4.18 Joules.

[1]According to the equation of state PV = ZRT, which relates the pressure P and the volume V to the temperature T and the compressibility factor Z at the pressure and temperature considered, and to the constant R.

...thermal energy

Everybody has at one point or another vigorously rubbed their hands together. Warming occurs because the mechanical energy used for rubbing has been transformed into heat, which is another form of energy, referred to as thermal energy. Heat is often obtained from fire (through the combustion of wood, coal, gas or oil), from electricity using the Joule effect (radiators, ovens, heating pads) or from the Sun, which warms the atmosphere on a daily basis and through it, everything else.

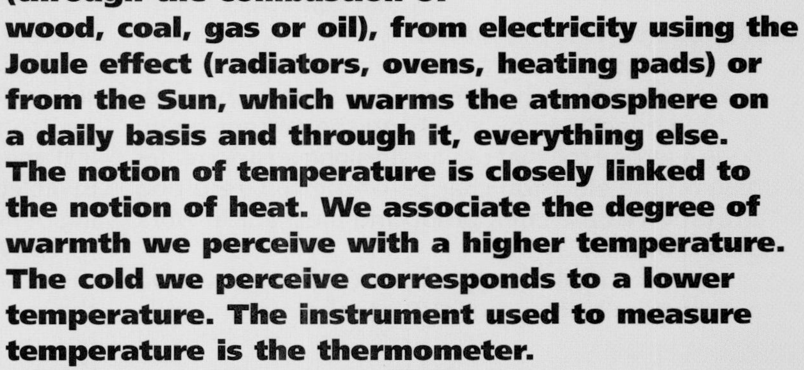

The notion of temperature is closely linked to the notion of heat. We associate the degree of warmth we perceive with a higher temperature. The cold we perceive corresponds to a lower temperature. The instrument used to measure temperature is the thermometer.

When we warm our hands in warm water or with a radiator, they do not move, and do not generate mechanical energy. Heat cannot be transformed directly into mechanical energy. In order to do so, a complex procedure is required: water is heated to a high temperature to obtain vapour, which is then expanded to activate a turbine. Such a transformation does not allow for total conversion of heat into mechanical energy: it has a limited efficiency. This is closely linked to the second law of thermodynamics, which will be discussed later.

A practical temperature scale has also been established, based, in addition to the triple point of water, on the normal boiling point of sulphur (717.75 K) and the normal fusion temperature of gold (1,336.15 K.), other stable and well defined properties. The intermediate temperatures below and between the first two fixed points are measured using a thermometer based on the thermal behavior of the resistance of platinum. Those between the last two are measured using a thermocouple of platinum/platinum doped with rhodium. Above the fusion point of gold, at very high temperatures, a total radiation pyrometer is used, an optical device that can determine the temperature of a body based on the radiative energy it emits.

Energy comes in all shapes...

Only known as mechanical or thermal energy in the 19th century, energy exists in many other forms.

Electrical energy

Electricity covers many phenomena associated with electrical charges in motion or at rest. At rest, these charges generate electric fields that can affect neighboring bodies; when in motion, they form an electrical current and produce a magnetic field.
In an electric engine, the interaction between the magnetic field generated by a coil and the current going through a second coil that can rotate along an axis creates a force on the electrical charges of the coil, resulting in a rotation of the later, producing mechanical work in the process.
Note that electrical energy cannot only be converted into mechanical energy, but also into thermal and radiant energy.

Radiant Energy

Radiation carries energy, even through the void of space. Radiant energy from the Sun is directly or indirectly responsible for most of the energy we use on Earth, yielding on average 1 kW per square meter in the form of visible light and infrared radiation.

A radiator transfers the heat produced by the Joule effect (heating of a conductor by an electric current) through infrared radiation. Radiant energy is also emitted by the filament of a light bulb through the same Joule effect. The low energy of radio waves can transmit information at a distance. A microwave oven heats food by causing vibrations of certain molecules through high frequency electromagnetic radiation generated from electricity. While radiant energy is often produced using electricity, it is important to consider that it can also come from other sources such as a flame. Any body at finite temperature emits radiant energy: the greater its temperature, the higher the intensity and the energy of the electromagnetic radiation it emits.

Chemical Energy

Chemical energy is associated with the binding of atoms in molecules. A chemical reaction generally causes a variation of the chemical energy of molecules, that is usually released and transformed into another form of energy, generally heat. Combustion involves the chemical reaction of a fuel with oxygen, which generates an amount of thermal energy equal to the difference between the chemical energy of the fuel, the oxygen consumed and the chemical energy of the combustion products (water vapour and CO_2).

Nuclear energy

Nuclear energy comes from the nuclei of atoms. The nuclei, 100,000 times smaller than the atoms themselves, are made of elementary particles – protons and neutrons – strongly bounded together. Similary chemical energy comes from the chemical binding of atoms in molecules, nuclear energy is associated with the binding of neutrons and protons into nuclei through nuclear forces. Nuclear energy scales are considerably larger than chemical energy. A nuclear reaction, by changing the structure of nuclei, can release a large amount of energy. In current nuclear power plants, fission reactions (involving the splitting of uranium nuclei into other nuclei about two times smaller in general) release a large quantity of heat, a third of which is converted into electricity by gas turbines and alternators.

Electromagnetic radiation is essentially a propagating electromagnetic field oscillating at a fixed frequency. Typical wavelengths range from the order of several meters to 10^{-11}m. In order of increasing wavelengths we have: the radio waves of radio and television; the wavelengths used by radar and microwaves; visible light, in-between infrared and ultraviolet; X-rays, gamma rays and finally cosmic rays.

...The diverse forms of energy

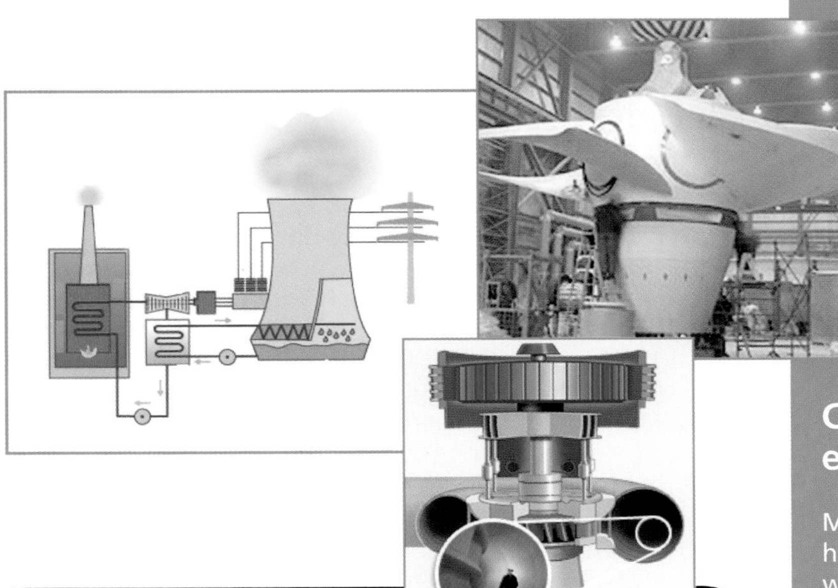

Nuclear fission

Free neutrons can split the uranium nucleus into two lighter elements such as bromine, lanthanum, krypton, barium, etc, generating a considerable amount of heat and freeing two or three neutrons, which can each provoke the fission of another uranium nucleus, resulting in a chain reaction. While the most common form of uranium (uranium 238) is only fissile under the action of fast neutrons, the nuclei of uranium 235 and plutonium 239 can be split by any neutrons, particularly if they are slow.

Nuclear fusion

Nuclear fusion consists in bombarding tritium atoms (isotopes of hydrogen with two neutrons) with protons with enough energy to fuse them into helium atoms, releasing in the process an amount of energy even greater than in fission. This reaction occurs in stars (such as the Sun) and is responsible for the radiant energy they emit. It is also used in thermonuclear bombs. It is extremely difficult to produce a controlled fusion reaction that could be used to generate usable energy. Fusion reactors are the object of several research activities, in particular of the ITER project, an experimental fusion reactor that will be built in Cadarache in France, through international cooperation.

$E = mc^2$

In both the cases of fission and fusion, the mass of the products of a nuclear reaction is slightly smaller than the mass of its initial components. This missing mass corresponds to the energy produced by these reactions, calculated according to Einstein's famous formula.

Chemical energy and electricity: energy to power engines

Mechanical energy has long been animal or human in origin, or obtained from the wind or waterways. Nowadays it is produced by thermal or electrical engines.
The mechanical energy generated by a combustion engine comes from the expansion of the gases resulting from the combustion of the fuel, which contain chemical energy.
The mechanical energy produced by electrical engines results from the electromagnetic properties of electric currents. Electricity can be transformed into mechanical energy, which is produced with generators, dynamos or alternators, whose operating principle is roughly the opposite of the electrical engine.

Nuclear energy

Nuclear energy is the considerable amount of energy generated from the fission of heavy atomic nuclei into lighter ones. A large part of this energy is thermal. In nuclear power plants it is transformed into mechanical energy by gas turbines.

Radiant energy or electromagnetic energy

The sun is perhaps the source of heat we are most familiar with. The sun does not transmit its heat directly, but emits electromagnetic energy (basically light). Radiant energy is another form of energy that our bodies and everything around us absorb as heat.

Energy is ever changing...

Energy can change form

Energy can be transferred from one system to another: an engine transfers the energy it produces to the wheels of a car, electricity produced in a plant powers various appliances, heat diffuses from a radiator to the ambient air and eventually dissipates in the environment. Energy can also change its nature: elastic energy can be transformed into kinetic energy (bow and arrow), brakes transform mechanical energy into heat, etc. Successive transformations are also possible, for instance chemical energy is transformed into electric energy in an accumulator, and this chemical energy can be restored by an electric current. The following table gives several examples of transformations.

Energy is always conserved

Thermodynamics is concerned with thermal exchanges. The first law of thermodynamics states that the energy contained in an isolated system (a system that does not interact with its environment) remains constant. Even when energy is transferred from one system to another or when it is transformed in various forms, there is never loss or gain of energy: energy is always conserved. The dissipation in the environment of some fraction of the energy undergoing a transformation or a transfer is referred to as an "energy loss". Thus a power plant transforms only part of its energy into electricity, the rest remains unused in general. This is discussed in more detail in the next section.

Energy can be measured

All forms of energy are measured using the same unit, expressing the fact that energy can be transformed from one form to another. The standard unit of energy is the Joule, defined as the work performed by a force of 1 Newton applied to an object undergoing a displacement of 1 meter in the direction of the force. The Joule corresponds to a small amount of energy: the kinetic energy of a mass of 1 kg moving at a velocity of 5 km/h or the amount of heat required to warm up 1 gram of water by 1/4 c of a degree. Thus, units such as the kiloJoule (kJ, equal to 1,000 Joule), the MegaJoule (MJ, representing 1 million Joule) and the gigaJoule (GJ, or 1,000 MegaJoule) are commonly used. In economics, energy is sometimes measured in terms of the ton equivalent oil (teo), which is the energy released by the combustion of a ton of crude oil. This corresponds to 44.6 GJ, when measured in terms of the higher heating value (assuming that the water produced by the combustion reaction is liquid), and to 42 GJ measured in terms of its lower heating value (without taking into account the heat of condensation of the water produced).

Initial form	Means of transformation	Final form
Mechanical	Friction, brakes	Thermal
	Alternators and dynamos	Electrical
	Reactions with compression	Chemical
	Breaking certain crystals	Radiant
Thermal	Vapour turbines	Mechanical
	Thermoelectric cells	Electrical
	Endothermic reactions	Chemical
	Incandescence	Radiant
Electrical	Electric engines	Mechanical
	Electrical resistances	Thermal
	Electrolysis, accumulators	Chemical
	Filaments of a bulb, fluorescent tubes, radio waves, TV, radar	Radiant
Chemical	Diesel engines, explosives	Mechanical
	Combustion, boilers	Thermal
	Batteries, accumulators	Electrical
	Phosphorescence	Radiant
Radiant	Solar sail in space applications	Mechanical
	Thermal solar collectors	Thermal
	Photovoltaic cells	Electrical
	Photography	Chemical
Nuclear	Reactors	Thermal Radiant

...but always conserved!

Power

Power is defined as the energy used (transmitted or transformed) per unit of time. It is a rate of energy measured using the Watt, which corresponds to 1 Joule/second. Because this unit is small, the kiloWatt (kW), equal to 1,000 Watts, is often used.
A practical unit of energy, the kiloWatt-hour (kWh) ,can be defined from the kW. This unit corresponds to the energy produced by a 1 kW system (for instance an electric radiator) during one hour. This system yields an amount of energy of 1,000 Joules (or 1 kJ) per second during 3,600 seconds. Thus, a kWh corresponds to 3,600 kJ.
The kWh remains a small unit, typical of the energy consumed on a domestic scale. For industrial processes, the MegaWatt-hour, (1,000 times larger) is used. On a global scale, the TeraWatt-hour, 1,000 times larger, is commonly used.

Energy loss, energy production

Energy is conserved for the physicist, but from the point of view of a user it flows through a system to eventually dissipate into the environment. For instance:
- Gas burning in a boiler produces heat which is used to warm water to a temperature of 80 °C. The water flows through radiators throughout a building, maintaining its temperature at 20 °C. The heat of the building then dissipates into the atmosphere.
- In an electric power plant, high temperature heat is transformed into electricity and low temperature heat, which is generally not retrieved, is lost to the environment. The electricity produced will be partly transformed into heat as it is transported in wires and used to power various appliances.
- In a vehicle, the energy from the fuel is dissipated as heat by aerodynamic and mechanical friction forces (through the transmission, the wheels and the brakes).

Only part of the energy used to power a system is really useful during the various transformations it undergoes. Even if it is conserved overall, it will eventually be changed into a form of limited usefulness. In order to become useful, oil must be extracted and refined; natural gas must be collected, uranium must be extracted, renewable energies (such as hydraulic, solar, wind, geothermal and biomass) must be harnessed. Energy must be produced.

Energy in all of its forms: mechanical, chemical, electrical, thermal, nuclear or electromagnetic is measured using the same unit: the Joule, the standard unit of energy in the International System of Units. Energy can be converted from one form to another using engines, burners, radiators, turbines, dynamos, alternators, collectors, etc

Power

To be able to spend a lot of energy in a short time implies having significant amount of power. In other words, power is the amount of energy per unit of time. A power of one Joule per second corresponds to one Watt.
Conversely, power used during a certain amount of time corresponds to energy spent. Thus, electrical energy is often measured in terms of kilowatt-hours (kWh).

Transforming thermal energy...

Even if the total energy is conserved during energy exchanges, the various forms of energy are not equivalent or equally usable, particularly when heat is transformed into mechanical energy. It is better to use a battery to power a watch than heating it on a stove, and no amount of boiling water will ever be able to melt metal. Certain energy transformations are irreversible. Heat is ever-present in our energy processes, for heating, of course, but also to produce almost all of our mechanical or electrical energy. How can heat be transformed into useful energy?

The Carnot principle

Thermal engines which are used for such transformations, are subject to the second law of thermodynamics, also known as Carnot's principle. The Carnot principle states that no cyclic monothermal transformation can produce work. Consider for example a brake, initially at ambient temperature. When used, the brake absorbs mechanical energy, raises its temperature, loses its heat to the environment and then returns to its initial temperature. Thus,the brake operates in a closed cycle. Because it has only exchanged heat with a single source (the atmosphere), the transformation is monothermal. Carnot's principle state that the brake cannot work in reverse: it cannot produce mechanical energy from heat.

Thermal engines

This postulate has fundamental consequences. In order for a thermal engine to produce work, it must exchange heat with at least two sources of heat at different temperatures. Specifically, the engine must retrieve heat from a heat source and release it to a cold sink. Thus, in a thermal power plant, water is subjected to a thermodynamic cycle during which it receives heat from the boiler (heat source), performs work on a turbine coupled to an alternator to generate electricity – and is sent to a condenser cooled by water from a river (cold sink), into which the heat dissipates. This water is then recycled in the boiler. The work performed by the turbine is the difference between the heat retrieved from the boiler and the heat lost to the condenser. In the internal combustion engine of a vehicle, the working fluid is the air, which is first heated by a combustion process (heat source), then used to generate work with the piston, before being evacuated into the atmosphere (cold sink), to be replaced by fresh air for a new cycle. A jet engine works basically in the same way. The mechanical energy retrieved is used to accelerate gases towards the back, producing thrust.
A thermal engine can only generate mechanical energy from heat if it can deliver part of it at a lower temperature.

Carnot's principle, efficiencies and degraded energy

The Carnot cycle is an ideal thermodynamic cycle during which a working fluid remains at constant temperature when it yields or retrieves heat, and does not exchange heat when it is brought from the heat source to the cold sink. Under such conditions, the Carnot efficiency η_c, defined as the ratio of the mechanical energy retrieved to the heat obtained from the heat source, is maximum and only depends on the absolute temperature of the two sources :

$$\eta_c = 1 - T_{cold} / T_{hot}$$

where temperature T is the thermodynamic temperature, corresponding to the absolute temperature discussed earlier.
If the efficiency of an ideal thermal engine is limited to a fundamental level by Carnot efficiency, in practice its value is even lower because the heat exchanges at the sources cannot be performed at constant temperature, and certain technical limits come into play. For instance, temperature differences are required to perform the heat transfers between the sources and the working fluid, and the engine receiving the work from the working fluid is subject to friction.
While energy can be completely transformed into heat (for instance through mechanical friction in a brake or through the Joule effect in a resistor), the reverse is not possible because of the second law of thermodynamics. This is why heat is sometimes referred to as a degraded form of energy.

Entropy

According to Carnot's principle, the energy produced by a thermal engine during a thermodynamic cycle operating between two temperature sources is equal to:

$$Q_{hot} - Q_{cold}$$

Its efficiency is given by:

$$(Q_{hot} - Q_{cold}) / Q_{hot} = 1 - T_{cold} / T_{hot}$$

Attributing a negative sign to Q_{cold} because it corresponds to heat lost by the system, the following expression is obtained:

$$Q_{hot}/T_{hot} + Q_{cold}/T_{cold} = 0$$

which characterizes this ideal cycle. The cycle is said to be reversible because it can be operated in reverse

...Carnot's principle

order under the same conditions without adding extra heat. In a real system, the efficiency is lower than the Carnot efficiency, and the equation above is negative instead of being strictly equal to zero. The cycle is then said to be irreversible, because additional heat must be provided to operate it in reverse.
This result can be generalized to cycles with any number of sources, the sum of the ratios Q/T of each source being always zero if the cycles are reversible. This means that if a system changes from a thermodynamic state to another (in an open thermodynamic cycle), this sum will not depend on the way the system changes, as long as each step is performed reversibly. This sum can be considered as the variation of a function which characterizes the state of the system between the beginning and the end of the thermodynamic process. This function is the entropy S, which is expressed as:

$$S = \int dQ/T$$

dQ being the quantity of heat given to the system during an infinitesimal reversible transformation performed at a temperature T.
The entropy of a system remains constant if it returns to its initial thermodynamic state through a reversible transformation. The entropy will increase if the transformation is irreversible as is the case for real systems, which cannot return to their initial state without an external inflow of energy. This result can be applied to the Universe: its entropy is always increasing since it will never be in states it has already gone through, because it is an isolated system with irreversible processes which cannot receive external heat.

It is often said that entropy is a measure of the degree of disorder of a system, in the sense that it measures what would be required from outside to bring the system back to its initial state. It can be shown that reducing disorder, for example, restoring a broken object to its initial state, sustaining a system in a given state or keeping a living being alive requires that energy be brought from outside the system considered. This implies that the entropy of the external system supplying this energy will increase. Thus, any system that maintains its level of organization will do so at the expense of an external system, raising in the process the entropy of the latter.

The transformation of heat into mechanical energy requires two heat sources at different temperatures. Its efficiency is proportional to the temperature difference between the sources. This constitutes the second law of thermodynamics (or Carnot's principle). The efficiency of this transformation is the lowest of the energy conversion processes, and the reason why heat is sometimes referred to as a degraded energy.

Sadi Carnot

Rudolf Clausius

Cooling machines and heat pumps

A cooling machine is an inverted thermal engine. Using a compressor powered by an electric engine, mechanical energy is supplied to a working fluid, which can then transfer heat from the cold source to the heat source. This process is how a refrigerator cools its content (cold source), transferring heat to the room where it is located (heat source).
A heat pump is a cooling machine used for heating. It takes heat from outside, which acts as a cold source, and transfers it inside a building at a much higher temperature, thus heating the building. The efficiency of the system is given by the ratio of the heat received to the electrical energy consumed. It is limited by the second law of thermodynamics, and is inversely proportional to Carnot efficiency. It is usually of the order of 5. The best heat pumps have efficiencies lower than 1. The relatively high cost of heat pumps and their complexity prevents their widespread use.

Annex 2

What will be the price of hydrogen at the gas station?

In the following section the cost of hydrogen produced from various sources will be estimated. However, these should be considered as rough estimates

In the following, a centralized production of hydrogen from fossil fuels is assumed, including CO_2 sequestration. The costs of storage, transportation and distribution are taken into account. Various studies suggest that a decentralized production scheme would be economically advantageous. This option is not considered here.

Hydrogen production

The following production modes of hydrogen are assumed:
- steam reforming of natural gas
- coal gasification
- biomass gasification
- water electrolysis

Table I shows the production capacities and the capital costs of these four technologies.
The syngas (H_2 + CO) obtained by reforming or by gasification is purified, usually through a PSA (Pressure Swing Adsorption) process, which produces a gas over 99.5 % pure (up to 99.9999 % vol.). As of yet, there are no specifications on the quality of hydrogen required for transportation applications, but the CO level must be below 10 ppm for a PEM fuel cell (purity is not a factor for an internal combustion engine). The level of purity required will influence the choice of the production method (centralized or local) and of the purification process (PSA, cryogenic separation, membrane separation), and thus the price of hydrogen. The higher the purity required, the lower the overall efficiency of the hydrogen delivered.
The production costs of hydrogen are given as a function of the price of the energy consumed by a given process. They also take into account the costs of the investments required and production costs per se.

Natural gas reforming

The SMR (Steam Methane Reforming) process is currently the most commonly used to produce hydrogen in large quantities.
The most important technology providers are:
- Technip-KTI (France),
- Foster Wheeler (US),
- Haldor Topsoe (Danemark),
- Howmar (US),
- Linde (Germany),
- Lurgi (Germany),
- Kellog (US)

The capital costs depend on production capacity, ranging from \$37 to \$42 for Nm³/day. For an optimal capacity (2.83 millions of Nm³/day or 250 t/day), the capital costs total 138 million dollars (Table I). The production cost of hydrogen will depend mostly on the price of natural gas. Assuming amortization over 20 years, and a return on investment of 15% before taxes, the price P_{H2} can be obtained from the price $P_{natural\ gas}$ using the following equation:

$$P_{H_2} = 1{,}275\ P_{natural\ gas} + 3{,}40\ (\$/GJ)^1$$

Assuming that the price of natural gas is 9.3 \$/GJ (January 2006), the price of hydrogen produced by a SMR unit is 15.3 \$/GJ. For reference, the cost of a litre of gasoline after refinement is about 18 \$/GJ.

Coal gasification

The Koppers-Totzek (Germany) and Texaco (United States) coal gasification processes are the most commonly used. For this estimate, the Texaco process, which operates at 70 bar and 1,200 °C, is used. This process, which uses large quantities of ashes, requires two to three times more capital investment than steam

[1] GJ : gigajoule, 1 GJ is 277 kWh or 0.023 tep (Ref: units and symbols).

methane reforming (see Table I). In addition, the efficiency is only 48 %, compared to 78.5 % for SMR.

The price P_{H2} of hydrogen is given, as a function of the price P_{coal} of coal by the following equation:

Table I. Production of hydrogen

Production mode	Production capacity	Capital costs (millions of $, 2006)
Steam reforming of natural gas	2.83 millions Nm³/day	138
Coal gasification	2.83 millions Nm³/day	490
Biomass gasification	1.31 million Nm³/day	288
Water electrolysis	2.83 millions Nm³/day	337

$$P_{H_2} = 2{,}08\ P_{coal} + 11{,}4\ (\$/GJ)$$

Assuming a price of 2.6 $/GJ for coal, we arrive at a price of 16.8 $/GJ for hydrogen.

Gasification of biomass

Biomass gasificators, currently under development for industrial use, operate either in a fixed bed or a fluidized bed configuration. The latter is more efficient and more flexible with respect to its feedstock.

The following estimate assumes that the Texaco process (with fluidized bed) is used. The biomass can either be dried and then fed to the gasificator, or pyrolyzed and processed as oil.

The production capacity is limited by availability (500,000 tons/year which corresponds to 50,000 hectares) and the cost of biomass close to the production unit. For a capacity less than half that of the SMR process, the capital cost is twice as expensive (see Table I). The price of the hydrogen produced as a function of the cost of biomass $P_{Biomass}$ is given by the following equation:

$$P_{H_2} = 1{,}73\ P_{Biomass} + 15{,}05\ (\$/GJ)$$

This leads to a cost of $ 19.5/GJ assuming a cost for biomass of $ 2.6/GJ.

Water electrolysis

Electrolysis, which has been in use for the production of hydrogen and oxygen since before the invention of SMR, remains the best process to obtain the very high quality hydrogen required by the semi-conductor and food industries. Using the currently commercially available processes, the cost P_{H2} of hydrogen produced by the electrolysis of water is given by the following equation as a function of the price of electricity ($P_{electricity}$):

$$P_{H2} = 1{,}25\ P_{electricity} + 7{,}10\ (\$/GJ)$$

Assuming a cost of electricity of $ 0.064/kWh (industrial cost of electricity in France in July 2004), the overall cost of hydrogen is $ 29.4/GJ for an electricity consumption corresponding to 52 kWh/kg of H_2. The price could be reduced by selling the oxygen produced ($ 35/ton), but a market would have to be found for the large quantities of oxygen produced. Even then the price of hydrogen would drop only by about 10 %.

Storage

In order to be used, hydrogen must be stored and transported either as compressed gas or as a liquid to the refueling station. Hydrogen can be stored on a large scale either in an underground system or as a cryogenic liquid.

Underground storage

This solution is the one currently used for natural gas. It would be the simplest to implement for hydrogen. It has already been realized by Gaz de France at Beynes, and in the salt mines of Teesside in the United Kingdom by ICI.

The overall cost of storage is a function of capacity and duration. Various studies show that this cost ranges from $ 1.7/GJ for a storage in a salt mine to $ 3.5/GJ for storage in a depleted natural gas reservoir.

Table II. Liquefaction of hydrogen

Capacity: 2.83.10⁶ Nm³/day – 255 tons/day – 85,000 tons/year	
Capital cost: 180 millions $ (2006)	
– Amortization (20 years)	8.5
– Return on investment (15 %)	27
– Electricity (12 kWh/kg H_2 at 0.064 $/kWh)	60.2
– Maintenance and labour costs	0.1
Total production costs (10^6 $)	95.8
Production costs ($/GJ liquefied H_2)	9.4

Storage as liquid hydrogen

Hydrogen must be liquefied at –253 C to be stored in liquid form. Liquefaction is an energy intensive process which requires at least 35 % of the energy content of hydrogen (12 kWh/kg H_2).
The cost of liquefaction (including capital investment and energy expenses) ranges from 8.5 to $ 9.5/GJ (Table II). The average cost of storing liquid hydrogen is about $ 5/GJ (Table V).

Transportation of hydrogen

Hydrogen can be transported by pipeline, train, boat or truck.

Pipeline

Hydrogen can be transported by pipeline as a pure gas or mixed with natural gas. In the latter case, the separation costs are significant. The price of the pipeline itself is a function of its diameter. There are currently pipeline networks linking distribution sites and industrial centres, spanning between 2,400 and 3,000 km, operating at various pressures (<100 bar). A classic 42 inch pipeline costs 1 million $/km. The cost of transportation per se is a function of capacity and distance. For large quantities of hydrogen delivered, this cost varies between 0.8 and $ 3.5/GJ. Thus for 800 km, a typical value would be $ 2/GJ if the pipeline operates at a pressure of 80 bar.

Transportation by rail, land or sea

Transportation by train is more attractive than by truck, both in terms of cost and the environment. For liquefied hydrogen, the transportation cost ranges between 0.5 and $ 1.25/GJ compared to 2 to $ 4/GJ for land transportation by truck (over distances ranging between 800 and 1,600 km). On the other hand, for compressed hydrogen, neither trains nor trucks can compete with pipelines: the transportation costs are more than 10 times higher ($ 20-40/GJ versus $ 2/GJ). The costs of transporting hydrogen by boat will depend on the distance, but remains more costly than pipeline: about $ 14-15/GJ.

Refueling station

The United States Department of Energy (US/DoE) has performed a study of a refueling station distributing hydrogen either as compressed gas at 600 bar, or in liquid form. The results of the study are shown in Table III.

Table III. Refueling station

	Compressed H_2	Liquefied H_2
• Number of vehicles per day	180	180
• Storage capacity of the tank (kgH_2)	6	6
• Average filling rate (kgH_2)	5	5
• Production capacity (kgH_2/day)	900	900
• Exit pressure (bar)	575	575
Investments ($ 2002)		
• 4 H_2 distributors of 1kg/minute for each pump	120,000	120,000
• 2 H_2 compressors (35 to 575 bar)	420,000	-
• Storage tanks (Compressed or liquefied H_2)	700,000	200,000
• Liquid H_2 pumps	-	80,000
Capital ($ 2002)	1,240,000	400,000
• Amortization (20 years)	62,700	19,350
• Return on investment (15 %)	188,100	58,050
• Operating costs ($)	357,000	307,000
Cost of distributing hydrogen at the refueling station($/GJ)	15.70	9.95

Table IV. Cost of hydrogen at the pump $/GJ

	compressed H_2	Liquefied H_2
• Steam methane reforming (at 9.30 $/GJ)	15.30	15.30
• H_2 liquefaction (assuming a cost of electricity of 0.064 $/kWh)	-	9.40
• Storage	1.70	5.00
• Transportation	2.00 *	1.25 **
• Distribution***	15.70	9.95
Net cost of H_2 ($/GJ)	34.70	40.90

* Pipeline. ** Rail + road. *** Including compression (3 kWh/kg of H_2).

Table V Net cost of compressed hydrogen at delivery as a function of feedstock and production process

Freedstock	Natural gas (steam methane reforming)		Coal (gasification)	Biomass (gasification)	Electricity (electrolysis)
Cost of the feedstock ($/GJ)	9.30		2.60	2.60	17.80
	Before CO_2 sequestration	After CO_2 sequestration	Before CO_2 sequestration		
• Production cost ($/GJ)	15.30	17.60	16.80	19.50	29.40
• Storage ($/GJ)	1.70	1.70	1.70	1.70	1.70
• Transportation ($/GJ)	2.00	2.00	2.00	2.00	2.00
• Distribution ($/GJ)	15.70	15.70	15.70	15.70	15.70
Net cost H_2 ($/GJ)	34.7	37.0	36.2	38.9	48.8

The refueling station uses tanks that can store 344 litres of hydrogen at 35 bar, corresponding to an overall amount of 1400 m³ or 4000 kg of hydrogen (corresponding to 4 days fuelling on average).
Hydrogen is then compressed from 35 to 575 bar to fill up allow at a rate of 1 kg per minute (fast-fill refueling). Based on these assumptions on the capital and operating costs, the cost of delivering hydrogen at the refueling station would be $ 10/GJ for liquefied hydrogen and $ 16/GJ for compressed hydrogen. The difference between the two is essentially due to storage (investment) and compression (cost of electricity). When the costs of liquefaction and storage are included (Table IV), compressed hydrogen is less expensive.

Cost of hydrogen

The cost of fossil fuels after refining ranges from $ 18/GJ to $ 70/GJ including the TIPP taxes. Table V shows that hydrogen costs between 35 and $ 50/GJ.
Hydrogen is thus 2 to 3 times more expensive than refined oil, but 1.5 to 2 times less expensive once taxes are included. The cost difference is mainly due to storage and conditioning. Note that electrolysis is the most expensive option.
Compressed hydrogen is only slightly less expensive than liquefied hydrogen (34.7 compared to 40.9), despite a high distribution cost, due to the price of compression and storage. Liquefaction is, however, less efficient because of its energy requirements (12 kWh/kg of H_2). The cost of sequestering geologically the CO_2 emitted during hydrogen production from hydrocarbon feedstock should be included in the overall price. With current technologies, the capital costs must be doubled to capture and sequester CO_2. Under current conditions, this would lead to an additional cost of $ 2.3/GJ of hydrogen produced, assuming that the cost of CO_2 capture is $ 44/ton.

Conclusion

The price of hydrogen for fuel cell or ICE vehicles is high because of costs associated with compression, storage and distribution, if a centralized production scheme is assumed. Under current conditions, the production of hydrogen by the electrolysis of water is not competitive compared to reforming, even taking into account the sequestration of CO_2.
On-board reforming of hydrocarbon liquid fuels has been considered by manufacturers to produce hydrogen for fuel cell vehicles. In addition to the technical challenges (purity of hydrogen, sizing, dynamic response of the system, start-up delay), this strategy leads to a technology that can no longer be considered "zero emission".
A possible alternative is to produce hydrogen locally at the station, which would eliminate the additional storage and transportation costs.

However, smaller scale hydrogen production units (reforming or electrolysers) are not yet as efficient as large scale systems. In addition, CO_2 sequestration is difficult for small units based on natural gas reforming. Improvements in productivity are always possible in the centralized production of hydrogen. For example:
- The production processes from fossil fuels and biomass can be improved, using, for instance, a combination of SMR and partial oxidation, or using oxygen instead of air by filtering with ceramic membranes for the preparation of syngas;
- Novel storage system could be developed using, for instance, adsorption or absorption storage in solids.

The price of hydrogen at the pump, excluding taxes, would range between the price of gasoline before and after taxes. Fiscal incentives, such as those currently available for clean fuels, could make hydrogen in gaseous or liquid form readily available to consumers. Finally, note that the present comparison has not taken into account the externalities associated with the impact of pollution by fossil fuels on public health and the environment.

Glossary

Adsorption: retention of the molecules of a gas at the surface of a solid due to solid-gas inter molecular forces. Adsorption depends on the nature of the gas and is a function of pressure. It can thus be used to separate and purify gas mixtures. Adsorption is a physical process and must not be confused with absorption, which involves chemical binding of the gas to the solid.

Atom: smallest particle of a chemical element. The atom is composed of a nucleus (protons and neutrons) and of electrons.

Biofuel: renewable alternative fuel from plants (ethanol and colza or sunflower esters, biogas).

Biochemistry: science of the chemistry of life (chemical structure of living beings and chemical phenomena associated with life).

Biogas: mixture of gases of biological origin, mainly composed of methane and carbon dioxide.

Biomass: living matter subsisting in equilibrium on a given surface of the earth. Biomass from vegetal matter is essentially a reserve of solar energy stored chemically through chlorophyll. Leaves act as cells that can convert solar energy into chemical energy. Solar energy stored as biomass can be converted back through thermochemic processes (combustion, carbonization, gasification) or through biochemical processes (methane or alcohol fermentation).

Cryogenic: Pertains to the technologies associated with very low temperatures, typically when most gases are in a liquid state.

Catalysis, catalyst: Catalysis is the action through which a substance (the catalyst) initiates a chemical reaction, or increases its rate, without actually taking part or being modified by the reaction.

Cogeneration: Simultaneous production of usable heat and electricity in a given system to improve its overall efficiency.

Digestor: Chemical reactor that can produce biogas from manure and by-products of water treatment plants using methanogenic bacteria.

Energy: A quantity that represents the capacity of a body or a system to produce work or raise temperature.

Eutectic: Pertains to solid mixtures which solidify at constant temperature, such as those of pure gases.

Energy vector: Form of energy that can produce work or heat but is not found in its usable state in nature. An energy vector can only be obtained by spending primary or renewable energy. An energy vector can be used to transport or store energy. Electricity and hydrogen are examples of energy vectors.

Fission and nuclear fusion: see nuclear energy.

Fuel cell: Electrochemical device in which hydrogen and oxygen are combined to form electricity, water and heat based on the reverse principle of electrolysis.

Geological times: periods of the history of the earth: primary, secondary, tertiary and quaternary era.

Geothermal energy: Basically heat from the earth's core. High temperature geothermal sources (100 to 300 C) can be used to produce electricity, whereas lower temperature sources (less than 100 C) can be used for heating.

Heat: Form of energy that can raise the temperature or induce state changes in a substance. Thermal energy is synonymous with heat.

Heating value (upper and lower): The heating value of a combustible material represents the quantity of heat that one unit of mass of this material can produce under normal conditions. The upper heating value refers to the case where the water produced by the combustion reaction is obtained as a liquid, whereas the lower value refers to water obtained as vapour. The difference is equal to the heat of condensation of water.

Heat pump: Thermodynamic system that can extract heat from a cold source and transfer it to a warm source through evaporation, compression and condensation of an appropriate fluid. The cycle is the same as for a refrigerator, but operated in reverse.

Heterotrophic: A heterotrophic organism is a life form that requires organic substances to grow or sustain it self.

Hydroelectric power plant: electric power plant that relies on hydraulic power from falls or the current of rivers to activate turboalternators to produce electricity.

Infrared radiation: Invisible electromagnetic radiation whose wavelength is immediately above that of visible red light, often used for heating.

Ion: An ion is an atom or a molecule with a net electric charge. A cation has a net positive charge, such as the ionized hydrogen atom, which is a proton (charge e+), or an atom of metal with valency n in a metallic solid, which has a charge of n e+. A negative ion or anion has a net negative charge. A substance which allows the motion of ions is an ionic conductor. Electrolytes are ionic conductors, which allow the motion of cations toward the cathode and of anions toward the anode under the action of an electric field.

Kerosene: Liquid fuel obtained from the distillation of crude oil used to power airplane reactors.

Metal hydrides: Hydrogen-metal compounds.

Molecule: The smallest particle of a body that can exist in a free state, generally formed of several atoms.

Natural gas: A mixture of gaseous hydrocarbons mainly composed of methane.

Nuclear energy: Thermal energy obtained from fission or fusion nuclear reactions. Fission usually refers to the splitting of the nucleus of the uranium 235 atom induced by collision with a neutron. This process also produces a large amount of heat and new neutrons. These new neutrons can in turn induce the splitting of other uranium nuclei, sustaining the fission reaction (chain reaction). Fusion is a nuclear reaction in which light atomic nuclei (such as deuterium and tritium) are transformed into heavier nuclei. This process also produces a large amount of energy, recoverable as heat. This reaction can only occur at temperatures of the order of hundreds of millions of Kelvins, and remain to this day a challenging experimental venture (ITER project).

Nuclear power plant: thermal power plant that relies on thermal energy from nuclear fission to produce electricity (see thermal power plant).

Oxidation: chemical reaction involving oxygen.

Photoelectric cell: device that relies on semiconductors (amorphous or crystalline silica, GaAs, SeS, etc) to convert electromagnetic radiation in the visible range into electric energy.

Primary energy: Energy that has not been transformed. Fossil energies such as coal, oil, natural gas, nuclear energy as well as solar energy are all primary energies.

Reforming, steam reforming: Chemical process through which a catalyst modifies the chemical composition of crude fossil fuels to increase their octane rating. Steam reforming is performed in the presence of water.

Renewable energies: Energy from inexhaustible natural sources such as the sun, wind, tides, water, the earth's core and biomass. Most come from the sun (except for geothermal energy which comes from the earth's core).

Supergeneration: See supergenerator.

Supergenerator: Fast neutron reactor which uses enriched uranium 235 or plutonium 239 to transform by neutron capture non fissile uranium 238 and thorium 232 into fissile isotopes, uranium 233 and plutonium 239. This reactor produces more fissile matter than it consumes.

Thermal power plant: electric power plant relying on the combustion of coal, oil, or natural gas to produce heat. The heat obtained is used to evaporate water in a boiler, which is then used to activate a turbine.

Thermochemistry: Branch of chemistry that pertains to the thermal aspects of chemical processes. By extension: the use of heat in chemical processes.

Tidal power plant: electric power plant that produces electricity using the energy of tides (turboalternators are set in motion by rising and falling tides which fill or empty a water basin).

Ultraviolet: Invisible electromagnetic radiation whose wavelength is immediately below that of visible violet light.

Wind turbine: A device designed to harness energy from the wind, consisting of a large propeller located at the top of a pylon. When rotating, the propeller activates a pump or an electricity generator. Wind turbines constitute the basic energy conversion device to convert wind power to electricity.

Abbreviations and acronyms

AFC Alkaline Fuel Cell (cf. § Fuel Cells).

ADEME Agence De l'Environnement et de la Maîtrise de l'Energie.

APU Auxiliary Power Unit, an electric power unit on-board the vehicles (most notably airplanes) used to power systems independently from the propulsion unit.

CEA Commissariat à l'énergie atomique.

CD Compact disc.

DVD Digital video disc.

DoE Department of Energy (USA)

EDF Electricité de France.

GDF Gaz de France.

HEC Hydrogen Engine Centre

HHV Higher Heating Value.

IEC International Electrical Commission

ISO International Standard Organisation.

ITER International Thermonuclear Experimental Reactor.

LHV Lower Heating Value.

LPG Liquified Petroleum Gas (a mixture of butane, propane from refineries, natural gas or oil reserves).

MCFC Molten Carbonate Fuel Cell (cf. § Fuel Cells).

NASA National Aeronautics and Space Administration.

NGV Natural Gas for Vehicles

PAFC Phosphoric Acid Fuel Cell (cf. § Fuel Cells).

PCRD Programme communautaire de recherche et développement.

PEMFC Proton Exchange Membrane Fuel Cell (cf. § Fuel Cells).

PSI Pound per Square Inch (cf. Units).

SOFC Solid Oxide Fuel Cell (cf. § Fuel Cells).

TEO Ton Equivalent Oil (cf. Units).

UNO United Nations Organisation.

USA United States of America.

Units and symbols

Ampere
The Ampere (symbol A) is the unit of measurement of the intensity of electric currents. An electric current of 1 Ampere is a constant electric current, which, when sustained in two parallel, straight conductors of infinite length with a negligible circular cross section, generates a force of $2\ 10^{-7}$ Newton per unit length between the two conductors when placed at a distance of 1 meter from one another.
The Ampere is the fourth of the fundamental units (the others being the meter, the kilogram and the second) which make up the MKSA international system of units, from which all other units used in mechanics and electromagnetism can be derived.

Ampere-hour
The Ampere-hour (Ah) is a measure of the quantity of electricity often used to quantify the charge of an accumulator. It corresponds to the amount of electricity carried in one hour by a current of one ampere.

Gram-atom, gram-molecule
The gram-atom is a unit that measures the quantity of an element whose mass expressed in grams is numerically equal to the atomic weight of the element. It is equal to the mass of the atom multiplied by Avogadro's number (6.023×10^{23}).
The gram-molecule (mol) is the quantity of matter corresponding to the sum of the gram-atoms of the elements constituting a molecule. A mole of hydrogen corresponds to 2.016 grams of hydrogen, a mole of oxygen to 32 grams of oxygen, and a mole of nitrogen to 28 g of nitrogen.

Bar
Unit of measurement of pressure (bar) corresponding to 10^5 pascals, roughly corresponding to atmospheric pressure (1 bar is equal to 0.987 atmosphere). The atmosphere is an old unit of pressure, corresponding to the pressure exerted by a column of mercury of 760 mm at 0 C at the standard acceleration of gravity (9.80665 m/s^2).

BTU
The British Thermal Unit (BTU) is a unit of heat equal to 1.05506 kJ corresponding to the heat required to raise the temperature of one pound of water by one degree Fahrenheit if its initial temperature is 39.2 F.

Calorie
The Calorie (cal) is a unit of heat equal to 4.184 Joules corresponding to the heat required to raise the temperature of one gram of water initially at 15 C by 1 C at atmospheric pressure.

Horse-power
The horse-power (hp) is a unit of power equal to 736 watts, frequently used in the automotive industry.

Degree Celsius or centigrade
The Celsius (C) is a unit of temperature such that the difference between the temperature of melting ice (0 C) and the boiling point of water at standard atmospheric pressure is 100 C.

Degree Fahrenheit
The Fahrenheit (F) is a unit of temperature such that it is the 180th part of the difference between the fusion temperature of ice (set to 32 F) and the boiling temperature of water at standard atmospheric pressure.

Degree Kelvin
The Kelvin (K) is a division of the absolute thermodynamic temperature scale defined in such a way that the triple point of water is 273.16 K. One Kelvin degree is equal to one Celsius degree. The absolute zero, on the Celsius scale, corresponds to -273.15 C.

Joule
The Joule (J) is a unit measuring the amount of three equivalent quantities: energy, work and heat. It corresponds to the work produced by a force of one Newton acting on an object over a distance of one meter along the direction of motion of the force. The multiples kilo, mega, giga and terajoules are commonly used to measure large quantities of energy.

Multiples
Multiples are prefixes corresponding to multiplicative factors of various units:

- Kilo (k) is a factor of 10^3=1000 (one thousand)
- Mega (M) is a factor of 10^6=1,000,000 (one million)
- Giga (G) is a factor of 10^9=1,000,000,000 (one billion)
- Tera (T) is a factor of 10^{12}=1,000,000,000,000 (one thousand billions)
- Peta (P) is a factor of 10^{15}=1,000,000,000,000,000 (one million billions)
- Exa (E) is a factor of 10^{18}=1,000,000,000,000,000,000 (one billion billions)

Newton
The Newton (N) is the unit measure of force in the International System of Units (MKSA). It corresponds to the force required to accelerate a mass of 1 kg at a rate of 1 m/s^2. The kilogram-force is a unit of force equal to 9.80665 Newton.

Normal cubic meter
The normal cubic meter (Nm^3) is a quantity of gas corresponding to the amount of gas contained in a volume of one cubic meter under standard pressure and temperature conditions (0.1 MPa and 273.15 K).

Pascal
The Pascal (Pa) is a unit of pressure of the MKSA system of units. One Pascal is the pressure exerted by a force of one Newton on a perpendicular planar surface with a surface of one square meter. One Pascal corresponds to 0.987×10^{-5} atmosphere.

Units and symbols
Acknowledgements

ppm
Short form of parts per million, corresponding to a factor of 0.000001.

PSI
Short form of Pounds per square inch, it is a unit of pressure corresponding to a weight of one pound exerted on a surface of one square inch, or 6,895 pascal (or 0.069 bar).

Ton equivalent oil (TEO)
The ton equivalent oil corresponds to an equivalent unit of energy used in statistical economics. One ton equivalent oil is the energy produced by the combustion of one ton of unrefined oil, or 44.6 GJ (HHV) or 42 GJ (LHV).

Volt
The volt (V) is a unit of measure of the difference in electrical potential (or tension). One volt represents the potential difference between two points when one Joule of energy must be provided to displace an electric charge of 1 coulomb from one point to another.

Watt
The Watt (W) is a unit of measure of power. It represents a rate of energy (mechanical, electrical or thermal) change of one joule per second.

Watt-hour, Kilowatt-hour
The watt hour (wh) and its multiples, particularly the kilowatt hour (kWh), are units of measure of energy commonly used to quantify electric energy. It corresponds to the energy provided by a power source of one watt over one hour, or 3600 joules.

Acknowledgements

We wish to thank the following organizations for the permission to use their photos in this book.

- Air Liquide, p.42, p.54, p.57
- Angstrom Power, p. 41 (top: charger and flashlight), p. 42 (helmet), p. 51
- Ballard Power Systems, p. 29 (fuel cell), p. 33 (bus), p. 40 (fuel cell), p. 42 (portable fuel cell, car)
- BWM, p.27 (car)
- Department of Energy/National Energy Laboratory, p. 46, p. 48 (wind turbines)
- Department of Energy/National Energy Laboratory and Nebraska Soybean Board (Soybean bus), p.19
- Department of Energy/National Energy Laboratory and Warren Gretz (switch grass, biomass in hand, gas plant, solar reflectors)
- European Commission, p. 37
- European Space Agency p.39
- Fuel Cell Technologies Inc, p. 31 (solid oxide fuel cell, top)
- Helion Fuel Cell Maker, p.41
- Hydrogen Engine Centre, p.27
- Hydrogenics Corporation, p. 29 (fuel cell illustration), p.31 (PEM fuel cell), p. 33 (small car, top), p. 42 (bus), p. 43, p. 47, p. 87
- Hydro-Québec, p. 13 (top), p. 19 (dam), P. 48-49 (dam), p. 77
- NASA, p. 53, p.55
- Ontario Power Generation Inc, p. 12
- Petro-Canada, p.13 (oil rig)
- Sandia National Laboratories, p. 13 (wind turbines), p. 19 (wind turbines and solar panels with hills),
- ThyssenKrupp Marine Systems AG, both pictures p.35

More about hydrogen

Books:

- "La révolution de l'hydrogène, *vers une énergie propre et performante ?*", Stephen Boucher, éd. Le Félin, Paris, 2005.
- " L'Energie de Demain ", J.L. Bobin, E. Huffer et H. Nifenecker, EDP Sciences, Paris, 2005.
- " Demain la Physique ", Alain Aspect, Edouard Brézin *et al.*, Editions Odile Jacob Sciences, Paris, 2004.
- "The Economic Dynamics of Fuel Cell Technologies", Arman Avadikyan, Patrick Cohend et Jean-Alain Héraud Editors, éd. Springer-Verlag, Berlin, 2003.
- "La course à l'hydrogène", Réal Godbout et Benoît Gauthier, éd. Soulières, Saint-Lambert, Québec, Canada, 2003.
- "The Hydrogen Economy", Jeremy Rifkin, Tarcher, New York, 2003.
- "Smelling Land" by David Sanborn Scott, Price-Patterson Ltd., Montréal, 2006.

Journals:

- International Journal of Hydrogen Energy, T. Nejat Veziroglu Editor-in-chief, Elsevier (available on line at www.sciencedirect.com).
- The Hydrogen & Fuel Cell Letter, monthly newsletter, Peter Hoffmann Editor and Publisher, New-York.

On the web:

- Mémento de l'hydrogène, a set of 62 technical notes on hydrogen energy prepared by the Association Française de l'Hydrogène (in French), www.afh2.org

Hydrogen Research and Industry

- Agence de l'Environnement et de la Maîtrise de l'Énergie, www.ademe.fr
- Air Liquide, www.airliquide.com
- Angstrom Power, www.angstrompower.com/
- Association Française de l'Hydrogène, www.afh2.org
- Association Lorraine pour la promotion de l'hydrogène, www.alphea.com
- Ballard Power Systems, www.ballard.com
- Canadian Hydrogen Association, www. h2.ca
- Commissariat à l'Energie Atomique, www.cea.fr
- European Hydrogen Association, www.h2euro.org
- Fuelcells Canada, http://www.fuelcellscanada.ca/
- Hydrogenics Corporation, www.hydrogenics.com
- Hydrogen Energy System Society of Japan, www.hess.jp
- Hydrogen Engine Center, www.hydrogenenginecenter.com
- Hydrogen Research Insitute, www.irh.uqtr.ca
- International Association of Hydrogen Energy, www.iahe.org
- International Energy Agency, www.iea.org
- International Partnership for Hydrogen Economy, www.iphe.net

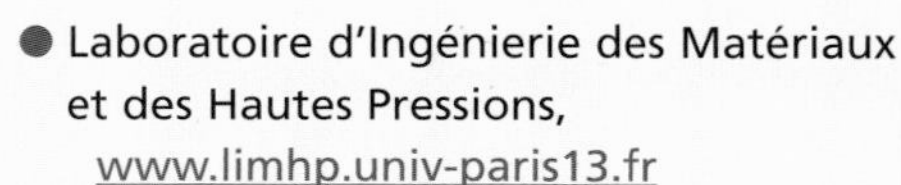

- Laboratoire d'Ingénierie des Matériaux et des Hautes Pressions, www.limhp.univ-paris13.fr

- Linde, www.linde.de
- National Hydrogen Association (USA), www.hydrogenassociation.org
- National Renewable Energy Laboratories, (USA) http://www.nrel.com

Achevé d'imprimer par Corlet, Imprimeur, S.A. - 14110 Condé-sur-Noireau
N° d'Imprimeur : 104543 - Dépôt légal : juin 2007 - *Imprimé en France*